Wild Work

Wild Work

Adventures of Women Field Biologists

Edited by
Susan B. Adams and
Sophie A. H. Osborn

SKIPSTONE

Published by Skipstone, an imprint of Mountaineers Books—an independent, nonprofit publisher. Skipstone and its colophon are registered trademarks of The Mountaineers organization.

Printed in China
29 28 27 26 1 2 3 4 5

Design: Jen Grable
Illustrations by Kira Cassidy, used with permission
Cover photographs, front cover, clockwise from top: *Kira Cassidy and arctic wolf on Ellesmere Island* (Photo by Ronan Donovan); *Zanethia Barnett prepares for a hookah dive in the Tennessee River* (Photo by Jeffrey T. Garner); *Lindsay Carlson records black brant nest locations and clutch sizes in Alaska's Yukon-Kuskokwim Delta* (Photo by Cody Deane); *Darcy Ogada holds a hooded vulture in Jaldesa Community Conservancy, Kenya* (Photo by Martin Odino).

In some stories, names have been changed and locations obscured to protect privacy. The views expressed in this book are solely the authors' and do not represent the views of the US Forest Service or other state or federal agencies.

Library of Congress record is available at https://lccn.loc.gov/2026001637.
Ebook record is available at https://lccn.loc.gov/2026001638.

Printed on FSC®-certified materials

MIX
Paper | Supporting responsible forestry
FSC www.fsc.org FSC® C188448

ISBN (paperback): 978-1-68051-877-1
ISBN (ebook): 978-1-68051-878-8

Skipstone
1001 SW Klickitat Way
Suite 201
Seattle, Washington 98134
206.223.6303
www.skipstonebooks.org
www.mountaineersbooks.org

LIVE LIFE. MAKE RIPPLES.

To my mom and dad, who always told me I could.
To the women who came before us and left the doors propped open after passing through.
And to the women who come after us, especially my daughter, Sophie Q, and my friend, Norah B.
—Susie

To field biologists, who are my heroes.
To the natural world that is my inspiration.
And to my sisters, Lisa and Natasha, who are my greatest champions.
—Sophie

Contents

CLOSE CALLS

HOPE AND HEARTBREAK

WILD FRIENDSHIPS

Preface

In April 2018, I (Susie Adams) walked into a large independent bookstore in Jackson, Mississippi, and made my way over to the nature section. As an avid reader and freshwater biologist, I was delighted to find a section devoted to aquatic books, with subjects ranging from fishing to identification guides to ecology—and even some fiction. But as I scanned the volumes, I was dismayed to discover that all the books appeared to be by or about men. I thought about a teenage girl I knew who loved wildlife, streams, and biology, and wondered, *Where are her role models? Where are the books showing young women that they, too, can become fish biologists?*

Every day, women biologists snorkel rivers to count endangered fish, hike through rough terrain tracking wolves and eagles, and scour remote landscapes for rare plants. But their stories are rarely told. I left that bookstore with a mission: to publish a collection of entertaining true stories that would show readers—especially young women and girls—that women biologists are thriving in challenging field biology careers. Such a book would demonstrate that outdoor careers in biology exist, are available to women, and can provide a lifetime of adventure and wonder.

The following year, I reconnected with a graduate school acquaintance—bird biologist Sophie Osborn, who had published a book on her work reintroducing endangered California condors to the wild and was working on a second book, *Feather Trails*. As we talked about our work experiences, I shared my book idea. Sophie's enthusiastic response and complementary skill set led us to join forces on the project.

Months later, we launched a request into our respective science worlds, asking for firsthand field stories by women biologists. The invitation spread like wildfire through biologist social networks. To our delight, stories poured in from women on several continents, though we had restricted our request to North American biologists to keep the project manageable. One of our most challenging initial tasks was choosing from among so many wonderful stories. Excluding submissions because they shared too many similarities or didn't quite meet the field story criterion, or because we simply didn't have room for them, was difficult. We wanted to share *all* of these amazing stories!

After whittling down the submissions, we spent several years working closely with the contributors to craft and refine their stories. Ultimately, we ended up with this delightful collection of riveting field stories that take readers behind the scenes, immersing them in the authors' work adventures and introducing them to the incredible creatures, plants, and places that inspire our commitment to nature and the outdoors.

Wild Work: Adventures of Women Field Biologists gives voice to the dedicated and courageous women who work tirelessly to study, protect, and conserve the world's wildlife and its habitats. Whether playing tag alongside wolves in the Arctic or wading among snakes in an Alabama river, evading poachers in Panama or surviving a flood in Alaska, being attacked by a bear or cuddling wild foxes, the thirty-seven women who lived these adventures and crafted these true stories have overcome fears, forged fierce friendships, and juggled fieldwork with family life as they embraced the natural world.

Field biologists of all stripes face a variety of impediments, both routine and unexpected. Some challenges, though, are unique to, or more commonly faced by, women. Perhaps the most obvious one is that peeing outdoors is more of a hassle for women! Other more serious challenges borne by women relate to men. Indeed, an overriding concern for many women biologists working in the field is not the danger presented by predators, snakes, or weather, but the potential threat posed by unwanted encounters with men. This risk can affect many decisions women make in the field, including where we park, how we camp, and which routes we take. And yet, despite the various dangers—actual or feared—that we face in the field, wild places often become our safe spaces, where we can recharge our batteries, escape society's pressures, and be among the animals, plants, and landscapes that intrigue and delight us.

Fieldwork can be physically demanding and even downright miserable, for both sexes. But since *some* physical tasks tend to be easier for men, women biologists often become adept at finding creative ways to complete such work. Alternatively, self-doubt or working with men who think women are not up to the job sometimes fuels the impressive strength and determination shown by many women biologists, who may even downplay serious fatigue or injuries to ensure they get the job done.

Many women also face discrimination and chauvinism during their science careers. Some of us have had to prove ourselves from scratch with each new job, while our male colleagues were assumed to be competent no matter their level of experience. For a subset of women, discrimination has included inequities in pay or opportunities, sexual harassment, or retaliation after resisting harassment by male colleagues or mentors. And at least in the US, women of color have faced even higher hurdles than white women have.

Despite encountering these and other challenges, women biologists worldwide have risen, again and again, to the demands of fieldwork—and thrived while doing so. Although some of the stories in this collection shed light on the unique challenges women face and the myriad creative paths they have found for overcoming them, this anthology focuses first and foremost on compelling field adventures.

Fears about the perils of fieldwork might lead some people (including family members, teachers, and advisors) to dissuade aspiring women biologists from embarking on such work, but most days in the field aren't marked by the sorts of dangerous or startling episodes described in some of these stories. Fieldwork is often mundane and uneventful, and our lives in the wild can be quite tame. Many of the adventures featured in this collection represent a single day or season out of years—or decades—of more routine fieldwork. But these unique experiences become our most often told tales. For anyone who might be fearful about what could befall your budding biologist, especially if a life spent in the wild is foreign to you, we urge you not to let your fears and anxieties impinge on your loved one's opportunity to experience the joys of fieldwork. Working in wild lands can be risky, but so can crossing the street, breathing toxic air in a city, or living a sedentary life. As you will see in this anthology, despite the occasional hazards, a career in field biology often leads to incredible fulfillment and a life of wonder and delight.

We invite you to hop in the truck—or boat or helicopter—to get an inside look at fascinating field days (and nights) in amazing places from the tropics to the Arctic. We hope you'll enjoy accompanying these adventurous women who reveal their strengths as they face their fears, cope with heartbreak, laugh out loud, and make lifelong friends while working in the wild.

—Susan B. Adams and Sophie A. H. Osborn

Rising to the Occasion

Finding Stella

Marta Curti

I sat in a makeshift blind in the middle of Belize's Chiquibul rainforest, trying my hardest to be absolutely still. Mosquitoes buzzed and bit. Sweat flowed freely, burning my eyes and prickling my back. To make matters worse, I really had to pee. But I was determined not to move. We were running out of time and resources, and failing again was not an option. This was literally a matter of life or death. Not for me or for our field crew. But for Stella.

Between 2002 and 2009, I worked for The Peregrine Fund's harpy eagle captive-breeding and reintroduction program in Panama, Central America. Measuring three feet tall from beak to tail, the harpy eagle is one of the largest and most powerful birds of prey in the world. Its strength lies mainly in its legs and feet, which are perfectly adapted to hunt the large arboreal mammals—such as monkeys and sloths—that live in Central and South America's rainforests. Despite its strength and size, the harpy eagle is at risk due to poaching and the destruction of its forest habitat. When The Peregrine Fund expanded its reintroduction project into Belize, I was sent to help prepare the new release site at the Las Cuevas Research Station, a facility established in the Maya Mountains for biological and anthropological investigations.

I got to spend three months there releasing captive-reared eagles to the wild with the help of some amazing volunteers. We cared for the birds before their release and tracked their movements once they were free. This project required long days in the field, bushwhacking through rainforest and climbing steep hills slick with mud and rain, all while resisting the urge to cling to tree limbs for support, since many were covered in needle-like spines or served as resting spots for perfectly camouflaged pit vipers, a group of venomous snakes with heat-detecting sensory pits on their heads. We tracked the eagles daily, stringing food in trees for them before they learned to hunt and rejoicing when we found young birds on their first kills—a signal that they would soon become independent and no longer rely on supplemental food. I loved all of it.

Though the forest in Belize was similar to the Panamanian forests I had become accustomed to, it felt different somehow. The trails, the landscape, even the smells were unfamiliar at first. A stone's throw from the Guatemala border, the Chiquibul Forest was full of lianas, vines as thick as my arm that grew in loops and swirls almost touching the forest floor. Poisonwood and chicle trees towered over us, the latter easily identified by the crosshatch scars on their bark, signs their sap had been harvested to make chewing gum. Spider monkeys chattered in the branches above us. Tapirs made their presence known only by the deep three-toed tracks we occasionally found along the muddy trails—some prints so fresh I was sure we had missed spying these shy animals by just minutes. Jaguars, too, roamed this forest, and the day one wandered into the field site—making eye contact with me before being swallowed by the dense vegetation—is a memory imprinted on my mind forever. The raucous flight of scarlet macaws added a splash of color to the palette of greens, hummingbirds flashed past with a buzz, and the watery calls of dark-winged oropendolas regaled us as we worked. Gray foxes, porcupines, coatimundis, ocelots, and a plethora of birds, reptiles, and insects called this place home. The forest was perfect.

Almost perfect.

Every once in a while, we came across telltale signs that reminded us that humans were in this forest, too: machete-cut paths, trash heaps, and even entire camps in disarray. Although most of the people who labored in the forest were just trying to make a living in a place with few resources and fewer choices, we heard stories of theft and violence as well. Sometimes, after dark, we even heard gunfire, apparently from battles between the Belizean army and individuals who were illegally harvesting xate palm leaves to sell to the floral industries in Europe and the United States. More alarming still was that some of these so-called *xateros* resorted to poaching wildlife to sustain themselves during their long ventures in the forest, something that could eventually prove hazardous to the eagles. Only a few months after our program began, incursions into the reintroduction area by poachers became more frequent, so we decided to relocate the three eagles we had been tracking to a safer expanse of forest in northern Belize.

My colleagues Angel Muela, Ryan Phillips, Phil Hannon, and Chris Hatten—some of the best field biologists I have ever worked with—and I quickly got to work trapping the eagles, which we were able to locate thanks to the three-and-a-half-ounce radio transmitters we had attached like a backpack to each bird before its release. In a short time, we caught and relocated two of the three eagles. Stella was the last one left to trap. As luck would have it, she had been spending all of her time within a few miles of the release site . . . until we decided to trap her, of course. Then, suddenly, she disappeared, along with the signal from her radio transmitter, which likely meant one of two things: either she had unexpectedly moved farther than she had ever gone before, or she had died—perhaps even been killed—and her transmitter had been buried, taken, or destroyed.

When you are trying to reintroduce a species, any species, every single individual counts. But when that species is a harpy eagle, a long-lived bird that takes years to reach sexual maturity, investments of time, energy, and emotion run high. Though it was inevitable that we would lose some eagles to human and natural predators—and we had—it didn't make it any less heartbreaking each time it happened. Our job was to do everything we possibly could to keep such losses to a minimum. We were determined not to lose Stella. So we spent days climbing the highest peaks we could find to improve our chance of hearing the signal emitted by her radio transmitter. Our bodies buzzed with hope each time we turned on the tracking receiver that would allow us to pick up her transmitter's signal, only to slump in disappointment when we heard just the high hum of static it produced when she was out of range.

Having failed to locate her radio signal from the ground, we arranged for an overflight in a small plane, which would allow us to avoid obstacles, such as hills and vegetation, that might block her signal. To our relief, once airborne we detected the clear *blip blip blip* of her transmitter about eleven miles from the field station and far from any roads. Back on the ground, we drove as close to her last known location as we could and then set out after her on foot. The going was rough and slow. We spent the entire day crossing rugged, rocky terrain, using machetes to whack through densely tangled underbrush. To make matters worse, the scarcity of fresh water in the area meant we had to carry all of the water needed for the journey on our backs. By nightfall, we were tired, sweaty, thirsty, hungry, and extremely disappointed to learn we hadn't even traveled two miles! We would never catch Stella this way. Though we were loath to admit it, it made more sense to wait for her to move to an area that we could access more easily. Accordingly, Ryan and Chris remained in Belize to monitor Stella's movements, while Phil headed home. Angel and I returned to Panama.

And waited.

A few tense weeks later, we got a phone call. Chris and Ryan had picked up Stella's radio signal in an area that was accessible by road. Angel and I quickly made plans for me to travel to Belize. However, by the time I arrived a few days later, Stella was already on the move and nowhere to be found. This time, after another overflight, we picked up her signal just east of the Guatemala border.

Wasting no time, Chris, Ryan, and I started packing. We filled our backpacks with tents, clothes, field gear, enough food for four or five nights of camping, as much water as we could carry, a trap, and ten pounds of meat and fur to use as bait. We piled our gear and ourselves into our truck and drove to Arenal, the closest road access to where we had last heard Stella's radio signal. We would be walking from there. Hoisting our heavy packs, we began the long trek into the forest. Nearly four miles later, as night was setting in, we made camp at the edge of a small cornfield.

The following morning, after a quick breakfast of oatmeal flavored with Tang (I don't recommend it), we headed out with radio receivers in hand. We searched for more than six hours. When I finally spotted Stella, high up in a tree, I couldn't believe how beautiful she looked, with her bright yellow feet, shiny black beak, and smooth, sleek feathers patterned with gray, black, and white. Now the question was: Would she be hungry enough to go after our bait?

I pulled the meat from my pack and held it over my head so she could get a good look at it. She immediately raised her crest and bobbed her head back and forth while staring at me intently—all signs that she was interested in the food. I felt a huge rush of relief. Without wasting another minute, Chris, Ryan, and I got to work setting up the trap and blind. We didn't have much daylight left and had to work quickly. We used rope to secure the small circular trap to a low horizontal branch in her line of sight. For this to work, she would have to come down to the bait and stand in just the right spot, and I would have to be quick enough to pull a rope that would tighten around her strong legs. Just as we were getting ready to bait the trap, we heard a strange call coming from right behind us. We froze. After a few seconds we heard it again. *Whooo, whoot*—the fake hooting call local poachers used to signal to each other.

Not wanting to take any chances by drawing attention to ourselves or to the harpy eagle perched in the trees overhead, we grabbed our packs and scrambled a few hundred feet up the mountainside, hoping the intruders wouldn't notice us, the bird, or the trap. Over the sound of my own heart hammering in my ears, I could hear Stella hopping from branch to branch. Willing her to be quiet, I kept my eyes and ears open. *If the poachers noticed her, would they try to hurt her? Could we react quickly enough? Did they know we were here? What would happen if they did?* After several tense moments, we heard their footfalls moving slowly away from us. The quiet that followed made it clear that the poachers had left the area, oblivious to the fact that they had just passed below one of the most amazing raptors in the Neotropics.

By the time we returned to the trap, night was setting in, and we knew Stella wasn't going to come down for food at this hour. Disappointed but determined, we set up camp nearby. After a quick dinner, I crawled into my tent and tried my best to sleep. Tree roots dug into my back, but I didn't have the energy to move my tent. Contorting my body into odd angles, I finally found a position that avoided the roots and was comfortable enough. But I couldn't quiet my mind. Though I knew Stella was safe for the night, I was anxious about what tomorrow might bring.

The next morning, we arose before first light, hoping to find Stella before she had time to capture her own prey. Luckily, we found her right away. We set the baited trap on a low perch, and I climbed into the dome-shaped blind we had built out of palm fronds and branches. Ryan and Chris moved several hundred feet away and crouched amid the vegetation, hidden from Stella's sharp eyes. From inside the blind, I could hear Stella flying from tree to tree just above me. About an hour later, she flew down to the bait we had placed at the center of the trap. This might be our only chance—we had to capture her.

My heart started pounding, and I was shaking so much I couldn't hold my hands steady. Even so, I tightened my grip around the rope. Seconds later, I pulled as hard as I could. I felt a tug at first, and then . . . nothing. The line went slack in my hands, and my ears flooded with the sound of disappointment—the *whoosh* of Stella's giant wings carrying her off into the forest. Peering through a small opening in the blind, my heart sank when my eyes confirmed what I already knew. The trap was empty. I had failed. And Stella was on the move again.

As we disassembled the trap, we were all quieter than usual, each lost in our own thoughts. Though Chris and Ryan did their best to remain positive, it was clear they were as disappointed as I was that Stella had eluded us. However, field biologists quickly learn to never give up. We were all the more determined to capture her—but we were swiftly running out of water. Fortunately, as if on cue, it started to rain early that afternoon, and the downpour lasted through the night. Though we, and much of our gear, were soaked by morning, being able to use the tent flaps to collect rainwater meant we could stay in the field at least one more day. One problem solved.

Our next problem was twofold. First, harpy eagles can easily go a week without eating. If Stella hunted successfully before we could find her again, it could be a long time before she was hungry enough to come down to our bait. Second, in the heat and humidity it wouldn't be long before the bait would become too rotten to interest her. If we didn't trap Stella today, we would have no choice but to go back to the field station empty-handed, possibly losing her forever.

That morning, we were up before sunrise. With gear packed, we headed to the top of a high slope to listen for Stella's radio signal. The heavy rain from the day before once again worked to our advantage, as it likely discouraged her from traveling too far. After only a few hours, we located her on the hill behind our camp. We set up the trap, and I hid inside another hastily built blind. This time, I was resolute that I would not miss. Just like the day before, I could hear Stella flying around above me. I willed myself to breathe and stay calm.

After what felt like an eternity, I watched from the gaps in the blind as she hopped to the branch and stood just a few feet from the trap. She glanced at the food, then lifted her head and froze. She was watching and listening intently. After a while, she took one tiny step toward the trap, lifted her head again, and paused. Over and over, she stopped to look around and listen as she inched closer and closer to the bait. All I could do was watch and wait.

Finally, she took one last look around and stepped onto the trap. This time, I took several deep breaths as she placed both feet onto the meat and began to feed. After each bite, she raised her head and looked around nervously, particularly toward the blind. I stiffened, hoping she wouldn't detect any movement from me. If I accidentally flushed her now, she might never come back.

After about ten minutes of me sitting stock-still and Stella cautiously eyeing the blind between bites, she turned her back to me and continued feeding. I gripped the line in my hand and waited for the right moment. Then, as she raised her head to swallow a piece of meat, I pulled the rope with as much force as I could muster. Again, I felt a tug and heard the *whoosh* of her wings as she tried to escape. But this time, I also felt the tension on the line. I didn't even have to look to know. We had caught her! I yelled for Chris and Ryan, who came running.

We approached Stella quickly and I immediately grabbed her legs to keep them under control. A raptor's first line of defense is its powerful feet and sharp talons. Harpy eagles have legs almost as thick as my wrists and rear talons the size of grizzly bear claws. Being grabbed by one can cause very serious injury and is to be avoided at all costs. Once I had a firm grip on her legs, I positioned her so that her back was against my chest. I used my arms to hold her wings in place—important for her safety and ours. A harpy eagle's wingspan measures around six feet but seems like more when attached to a struggling eagle. Once Stella was secure in my arms, Ryan placed a hood, specially designed for raptors, over her head, and I felt her relax right away. Not being able to see what was happening around her lowered her stress level. With Stella now safe and calm, Ryan and Chris set about dismantling the trap and blind.

We began the six-mile trek out of the forest, carrying Stella and all of our gear. We took turns holding her. Carrying fifteen pounds of the most powerful eagle in the world while hiking through a tangled forest on uneven terrain is not an easy task, and we couldn't risk anyone getting fatigued. The going was slow, and night set in while we were still more than a mile from our vehicle. Luckily, a full moon illuminated our way enough for us to safely navigate through the forest to our parking spot. Roughly four hours after beginning our return trek, we arrived at the car, exhausted but extremely happy. Stella struggled briefly as we removed her hood and gently placed her in a large pet carrier, but she quickly settled down. We then drove several hours to the Rio Bravo field station.

The next day, we took Stella out of the kennel, then removed her old backpack transmitter and outfitted her with two new ones—one we would use to follow her from the ground and the other to track her movements remotely, by orbiting satellites. With mixed feelings of relief and worry, we released her into her new home.

I, too, was headed back home—to Panama.

A few days later, I received word from Ryan and Chris. They had tracked Stella and found her feeding on a coatimundi. This was the best news I could have received. It filled me with hope that she would make a good life for herself there in the Programme for Belize's Rio Bravo Conservation and Management Area, one of the largest protected areas in the country. Is it a perfect forest? No. Sadly, nowhere is perfect for harpy eagles anymore. But this area would give her the best chance she would have to survive.

Even though years have passed since I last had word of Stella, I find myself thinking about her from time to time. And when I do, I like to believe she is still out there—chasing monkeys, joining the forest's chorus with her sharp calls, and taking her rightful place as an apex predator high up in the forest canopy.

Marta Curti was born and raised in Los Angeles. She grew up where coyotes and owls called at night and deer roamed the hillsides, leading to her love of all things wild. Also influenced by art and language, Marta majored in creative writing before earning a master's in education. She dreamed of becoming a teacher, but a post-college road trip spent camping and backpacking across the US convinced her that she wanted to work outdoors, conserving

Angel Muela

wildlife and wild places and instilling a love for the outdoors in others.

For more than two decades, she has worked for The Peregrine Fund, combining her love of wildlife, education, and writing to help conserve some of the world's most amazing raptors. As the organization's Hispaniola program director, she helps conserve the critically endangered Ridgway's hawk. She co-owns Whitehawk Birding and Conservation, an ecotour company focused on wildlife observation and environmental education, and has written five bilingual children's books and a novel, *In All Things*. Marta loves to travel, hike, camp, read, practice yoga, scuba dive, and otherwise enjoy nature.

Bison Rescue

Lisa Osborn Koitzsch

The ice beneath my feet cracked as I bent down next to the jagged opening and grabbed the bison's stubby horns with my cold, wet hands. Her soft, dark eyes rolled wildly, the whites communicating her fear and looming surrender. My heart pounded. *Come on, Ky!* I pleaded silently as I shared a distraught look with Molly, who was anxiously watching from the rocky beach. We desperately needed additional help. I shifted my position, hoping the ice beneath me was strong enough to hold me, as I secured my elbows against my knees and tried to keep the brown wire-haired head above the water. This was all my fault.

I was spending my ninth winter working in Yellowstone's Northern Range investigating clustered GPS (Global Positioning System) locations—points on the landscape where a radio-collared wolf and its packmates had spent more than an hour. By visiting these GPS clusters, we hoped to find wolf kills and then necropsy what remained of the carcasses to learn what prey wolves consumed, how much they ate, and why they targeted a particular individual over others. I was co-leading a crew of five, and we did all our work on skis—the most efficient way to travel across the rugged snow-covered landscape.

Molly and I had set out on this day in early March to search GPS clusters from a female wolf collared just the year before. After donning our ski gear, ensuring we had the tools we needed, and shouldering our field packs in the parking area closest to our destination, we skied downhill, moving carefully on the crusted snow and dodging the sagebrush that dotted the undulating terrain, until we reached the southern bank of a large creek. We continued along the rim of the steep bank for a little while, enjoying the 45-degree day and savoring the sunshine that poured over the open landscape. Despite the temperature, the broad creek was still mostly ice covered, although patches of open water were beginning to appear. Wildlife trails crisscrossed the snowy expanse. Bison had made the most defined paths, compacting the snow that covered the ice and creating highways these large animals used all winter.

As we neared an oxbow bend in the meandering watercourse, Molly and I spotted a small herd of bison cows and calves about a quarter of a mile away from us. All of them had their heads down and were pushing the deep snow from side to side—a behavior we called buffling—as they intently searched for tiny blades of green grass emerging under the snow. We slowed our pace, conscious of the scraping and crunching noises our skis made on the crusty snow, and cautiously began to pass the herd. One of the cows lifted her head and stared at us for several minutes. I held my breath and whispered, "Please don't run, please don't run!" Since the bison were surrounded on three sides by the sinuous creek and we were in their way on the fourth side, I worried they would try to avoid us by crossing the thinning ice.

Changing course, we angled away from the oxbow, hoping the cow would realize we didn't pose a threat and would go back to eating. No such luck. Her narrow tail slowly lifted—a sure sign of alarm in a bison—and with a snort, she whirled and began to trot away toward the creek. Instantly, all the other bison heads surged up—along with their tails—and the herd turned to follow her. Most of them seemed unaware of what they were running from, as they thundered toward a spot on the creek bank where they had clearly crossed the ice many times before. Instead of calmly crossing one by one on the existing bison trail, they jostled and scrambled across the wide ice-covered creek in a confused mass. In their mad dash, most of them made it safely to the opposite bank, but five bison crashed through the ice and into the frigid creek. Water and ice churned under the panicked animals. Thankfully, within minutes, four of them heaved their bodies up onto thicker ice and lumbered toward the shore. Water droplets flew as the bison shook themselves from head to toe and joined the rest of their small herd on the opposite bank. Some loped away through the sagebrush, but a few turned and peered back at the creek.

One lone bison remained in the water. A calf—born the previous summer and now weighing about four hundred pounds—flailed and splashed, trying desperately to haul her body up out of the hole she'd fallen through and back onto the ice. Frantic, panicked, and breathing hard, she swam from one side of the icy opening to the other. Her chest bumped against the edge of the ice and her hooves lashed out as she tried to gain purchase on the slippery surface. Molly and I watched in horror through our binoculars, hearts in our throats, willing the calf to pull herself out of the icy water. Five minutes went by, then ten. We looked at each other in dismay, painfully aware that our presence had caused the animals to panic. Now one of them was trapped—an all-too-familiar occurrence in Yellowstone, where bison often fall through the weakening ice on ponds and creeks in early spring. Some of the animals are lucky enough to make it out; others, too tired to fight after a long struggle, give up and drown. When the ice melts, their bloated carcasses provide food for a host of scavengers, including bears emerging from their winter dens.

Yellowstone National Park's official policy is to let nature take its course and to not interfere in natural events. I had witnessed many such dramas in the Yellowstone area, where helping to rescue an animal or putting it out of its misery might have been the most humane option, but because I was used to the park's justifiably strict noninterference policy, I'd always refrained from doing anything. Not this time. I felt like we had to do something to help the trapped calf since we had inadvertently caused this situation. But we were going to need some assistance.

Pulling my radio out of my pack, I pressed the transmission button and said: "Ky, this is Lisa. Do you copy?"

The reply came back almost immediately. "I copy. Go ahead."

Just hearing my husband and longtime field partner's voice was reassuring. I hadn't been sure he would be available since he was several miles away talking with a group of students about our work, but I knew I could count on him to help if he could.

"Please come and meet Molly and me . . . now." I spoke calmly as I gave him the coordinates for our location, but he recognized immediately that something was wrong. I was usually more verbose over the radio, something he often teased me about.

"On my way," he replied without any hesitation. He would have to drive two miles to the parking area and ski more than a mile to reach us.

Help was now forthcoming, but I also needed to contact one of our supervisors to let someone know what we were planning to do. Luckily, I had cell service and was able to reach one of the project leaders, who was also a trusted friend. I quickly summarized our situation and told him we wanted to try to get the bison out. He emphasized the importance of not interfering but then added, "Just be careful, and keep me posted."

The calf's mother had returned to the creek bank and was trotting up and down, grunting in distress, stamping her feet, and staring at us. Relieved that she was on the other side of the creek from us, Molly and I sprang into action. We picked up several large pieces of driftwood lying along the shore and proceeded slowly toward the trapped bison, walking gingerly out onto the ice, testing every step.

Once we reached the edge of the hole the calf had made when she fell into the river, we began hammering at the ice with our driftwood to try to create a larger opening and possibly open a path through the ice to the shore. I stood on the well-worn bison path that crossed the ice and pounded the frozen surface as hard as I could, feeling my whole body vibrate with each blow. *Bang, bang, bang.* I kept my eyes on the bison, hoping I wasn't making her situation even worse. She had moved as far away from us as she could get and was pushing against the ice and breathing hard.

As I stopped to give my arms, hands, and back a rest, Molly circled carefully around me to try to work on the edges of the ice from another side of the hole. Suddenly, I heard a loud crack followed by a splash. Whirling around, I froze in shock as I saw to my horror that Molly had fallen through the ice. Not daring to move for fear of breaking through as well, I watched as Molly, who was remarkably calm given how frigid the slow-moving water was, quickly pulled her athletic frame out of the river. In no time, she was sitting on the rocky shore, pulling dry clothes out of her pack and changing out of her wet ones. Thankfully, the sun was still shining and the temperature had increased slightly.

Once I was sure Molly was safe, I turned my attention back to the bison. She was still struggling, but her movements seemed slower and weaker, and her eyelids kept closing over her dark, frightened eyes. I continued to try to break the surrounding ice—to no avail. As I watched the calf, I realized that her head was sinking lower and lower until the water was almost lapping over her broad muzzle. Her breath pushed out in bursts, clouding the air.

Squatting down carefully, I grabbed onto her horns to bring her head between my knees and within a few feet of my face. Having given one of the park's most dangerous animals a wide berth for years, I could scarcely believe that I was now holding on to a bison's horns, hoping she wouldn't pull me in or batter me with her powerful head. Wary of her strength, I nonetheless marveled at how beautiful she was and how long her eyelashes were. Keeping her muzzle and head above water, I murmured softly to her, all the while desperately hoping the ice wouldn't break beneath my feet.

For what felt like an eternity, I clung tightly to the helpless calf's rough horns to prevent her from sinking below the water's surface. Finally, hearing a shout, I looked up and to my great relief saw Ky kicking off his skis and climbing down the small hill toward us. Molly quickly filled him in on what had happened. Before turning my attention back to the calf, I watched Ky pause for a few seconds as he took in the splintered ice, the bison cow pacing the far bank, and me crouching while holding a wild bison by its horns. Then, dropping his pack on the creek bank, he quickly stripped down to his long underwear, perhaps realizing he might end up in the water as Mollie had, and put his ski boots back on.

Not knowing what he was going to find when he reached us, he'd had the forethought to grab a long rope from his work truck before skiing out to meet us. Wasting no time, he made his way across the ice to where I was holding the calf's head and looped the rope around her horns. I let her go and moved aside. Molly joined me as Ky used the rope to guide the bison toward the side of the jagged hole closest to the shore where the calf's mother was frantically pacing. He then handed the end of the rope to Molly and me, and the two of us tried to slowly pull the bison up onto the ice, but the creek's slow current pushed her front legs under the frozen shelf in front of her. Exhausted, the calf made no effort to heave herself out.

While Molly and I kept tension on the rope, Ky knelt next to the bison and tried to lift one of her front legs up onto the ice. But with a crash, the ice broke beneath him, and he plunged chest deep into the icy water. Not hesitating for a moment, he lunged over to the large animal that shared the ice hole with him. Grabbing on to the horn closest to him with one hand to prevent her from goring him, he used his other hand to lift the bison's front feet, one by one, up onto the ice. Meanwhile, Molly and I pulled steadily on the rope that was around the calf's horns to keep her from thrashing in panic as she tried to move away from us. Every muscle in my body strained as the bison jerked her head backward, trying to escape us.

As soon as her front feet were out of the water and on the ice, Ky dove underwater, disappearing below the surface, and put his shoulder under the bison's hind end. Digging into the sediment at the bottom of the river with his ski-booted feet, he pushed upward with all his strength. With Molly and I pulling with all our might and Ky pushing, the bison seemed to pop effortlessly out of the hole. Within seconds, she was standing on the ice.

Molly and I looked at each other in stunned relief. But as soon as the bedraggled animal's back feet were out of the icy water and on a firm surface, she began bucking and whirling around. Hanging on to the rope for dear life, we struggled to keep our footing on the ice, not wanting to let go of the bison and risk her running off with a rope around her horns. But within thirty seconds, the ice gave way beneath her again and she fell back into the creek. Fortunately though, only her back feet broke through this time—and she was now in shallower water much closer to shore and to her anxious mother.

Ky clambered out of the water and ran toward us, urging us to keep holding the rope but to release some tension so that he could remove it from around the bison's horns. Working quickly, he slipped off the rope. As soon as the calf was free, Ky, Molly, and I hurried back across the creek to give the calf space to get out of the water on her own. Finally feeling solid ground underneath her back feet, the calf surged out of the water onto the ice, then leaped up over the snowy creek bank and onto the sagebrush flats where her fretful mother was waiting.

Once she was out of the water and off the ice, the calf stood for a few moments stunned, head hanging down, sides heaving. We held our breath, watching to see what she would do. Looking up, she stared balefully at us for a few seconds and then shook herself violently—water, mud, and ice flying off her thick coat. Finally, she whirled around, kicked up her heels, and ran toward her mother and the other bison from her herd.

Our relief palpable, smiles splitting our faces, we gave each other wet and muddy hugs and high fives. As wildlife biologists, we knew better than to interfere with nature and Yellowstone's iconic wildlife, but when we were the ones responsible for placing an animal in danger, none of us had hesitated for a moment to grab a bison by the horns.

Ky B. Koitzsch

Born and raised in Switzerland, Lisa Osborn Koitzsch moved with her family to Vermont when she was twelve years old. She earned a bachelor's degree in humanities and French literature from Johns Hopkins University and worked for several mindless years in administration and business directory publishing. Inspired by her sister Sophie's passion for birds and introduced to the conservation community by her sister Natasha, Lisa subsequently worked for a wildlife conservation organization in Washington, DC. This experience fueled a lifelong interest in wildlife, which led to work on numerous field research projects.

Lisa studied warblers in Vermont and helped analyze American dipper habitat in Montana. She tracked the habitat use of migrating songbirds in Belize and inventoried breeding birds in Costa Rica. While doing graduate research at the University of Vermont on how reforestation impacts grassland birds, she met her husband, Ky, a wildlife biologist and builder. Together, they owned a timber-frame building company for many years. During the winters, they studied wolves and moose in Yellowstone and Isle Royale National Parks. They currently live in Vermont's Mad River Valley with their intrepid rescue dog, Phoebe.

How to Hold a Vulture While in the Jaws of a Jackal

Darcy Ogada

"Show me where the jackal bit you!" my six-year-old gleefully demanded upon my return from a week in the field. No hugs, no "I missed you"—just "Show me your wound!" Clearly, he watches too many reality TV shows, and I was just one more victim with a story to tell.

My work as a biologist for The Peregrine Fund's Africa program mostly involves the conservation of African vultures. The biggest threat to these magnificent birds is poisoning, with vultures being indirect casualties of livestock owners trying to kill predators, such as lions and hyenas, with poison-laced carcasses. Unfortunately, vultures—nature's most efficient cleanup crew—often arrive at such carcasses by the tens, if not hundreds, and unwittingly ingest poison when feeding, only to perish along with the predators. As part of my work, we trap and tag vultures by attaching a new generation of GPS-GSM (Global System for Mobile Communications) transmitters (think miniature smart phones that are topped by solar panels, weigh less than two ounces, and fit easily into the palm of your hand) to the birds' backs to better understand their movements and hence their exposure to poisoned carcasses across East Africa's vast landscapes.

How does a person catch a vulture? you might ask. It requires the carving skills of a butcher, the knot-tying expertise of a sailor, and the patience of a saint. First you need bait, hence the butchery skills. A butchered goat is both a tasty scavenger meal and a vessel for vulture entrapment. Next, you knot a long piece of tennis-racket string to create a series of nooses. A line of eight or so nooses, one of which you hope will entangle a vulture's long toes as the bird feeds, constitutes the trap. To secure the trap to the goat carcass, you weave the nooses in and out of the rib cage and around the various appendages. The heavy weight of the carcass anchors the noose trap to terra firma. You don't want your noosed vulture flying away still attached to the bait. After setting the trap, you drive off to view the carcass from afar . . . and wait.

Despite their reputation as gluttons, vultures are cagey targets, even for experienced trappers. Our quarry, a Rüppell's vulture—one of the world's widest-ranging birds—finally arrived at the carcass. Alone. When it flapped its wings, I knew we had our bird. Jamming the car into gear, we arrived in less than a minute. Unfortunately, a jackal arrived even faster. And no sooner did it arrive, than it too got tangled in the trap as it attempted to dine and dash. Jackal and vulture were cinched on opposite ends of the trap line, teeth against beak.

Noosing a jackal was not part of our plan. Nor was the ensuing melee among feathered, furred, and bare-skinned beasts. Leaping from the car, my trapping partner, Simon, yelled, "You get the vulture, I'll get the jackal." I jumped into the fray, but I'm no Muhammad Ali, and this was one rumble in the jungle where being bigger certainly didn't end up being better. I was outfoxed by a fox. Or rather, a jackal.

Jackals are frequently uninvited guests at our traps, so Simon was prepared. One arm ensconced in a thick leather glove that reached to his elbow, he attempted to entice our unwanted guest toward freedom. Kneeling by the carcass, I quickly settled into mother hen mode, grasping the vulture by its powerful neck and supporting its body—still bound to the carcass—as I scooped it into my lap, at which point the jackal morphed into comic cartoon villain Wile E. Coyote. Picture an oversized, caffeine-laced Jack Russell terrier unable to break free from its leash. My allegiance being firmly with my entangled feathered friend meant taking one for the team. And I did—on my left hip.

Perhaps the bite was a release of pent-up jackal aggression. Let's face it, it can't be easy playing second fiddle to a host of sexy alpha predators—lions, leopards, and cheetahs—strutting their stuff on *Big Cat Diary*. With the jackal soon subdued and released, my mind flashed to what jackals are most notorious for—carrying rabies.

But before I could start foaming at the mouth, we had a vulture to attend to. Fortunately, the bird proved to be as cooperative as the jackal was maniacal and calmly reclined in my lap while we attached the lightweight transmitter to his back and quickly sent him on his way.

Freely roaming the skies again, our tagged vulture showed us habitats that the species uses for nesting and foraging. Obtaining such information is essential to conserving these spectacular but imperiled scavengers. And while saving the vultures may mean feeding a few jackals along the way, preferably next time it will be with our bait rather than my behind.

Franco Ogada

Growing up in a sporty and adventurous family in upstate New York, Darcy Ogada embraced the outdoors. A solo springtime camping trip to Yellowstone National Park was the catalyst for her journey into wildlife conservation. After completing a two-year degree in forestry from the New York State Ranger School, she joined the Peace Corps in Niger, West Africa, where she helped communities establish tree nurseries and a seed bank. She went to Kenya to conduct bird research for her master's degree and never left. Along the way, she had a son and completed her PhD at Rhodes University in South Africa and is now a dual Kenyan-American citizen.

As the Northeast Africa program director for The Peregrine Fund, she has helped expose the widespread poisoning and severe decline of Africa's vultures. The unceasing pressure facing wildlife in the developing world overwhelms her, but she awakens each day maintaining the deranged belief that she can still save the planet.

The Flood

Lindsay Gray Carlson

In spring, I usually woke to the *kiau* and *cow-cow-cow* of glaucous gulls circling over my tent as though I were a french fry in a mall parking lot. I was usually jostled out of sleep by the *kuk-kuk*, *oo-ou*, and guttural croaks of tundra swan courtship duets reverberating across the mudflat opposite my cot. These two white-feathered species were always among the first birds to arrive each April on western Alaska's still snow-and-ice-covered Yukon-Kuskokwim (Y-K) Delta. But on *this* morning, April 29, I woke instead to the clamor of a flapping tent fly, the roof billowing as if it were about to lift off, the seams stretching nearly to the breaking point. Outside, the blades of our wind turbine created an almost hypersonic whirr. Strong gales are a familiar foe in coastal Alaska, but the absence of bird calls should have been a warning of worse weather to come.

It was my fourth time witnessing spring's arrival in the Yukon Delta National Wildlife Refuge. I had come for a summer of field science and solitude, but things were off to an unusual start because the previous winter had been an especially problematic one. The typical high-pressure, sunny-but-frigid subarctic weather systems had been replaced by heat-trapping low-pressure systems more reminiscent of a bad Vancouver ski season. A sustained lack of shorefast ice had prevented Yup'ik and Cup'ik hunters from seal hunting for most of the winter. The unusual weather conditions and record low ice extent were partially a result of the aptly named Blob, a large mass of warmer-than-usual water in the North Pacific—an anomalous pattern classified as a marine heat wave.

When the field camp was established more than twenty years earlier, the site was selected for easy access to a particularly productive goose nesting colony. The camp, sitting on the inside of a horseshoe bend in a tidal river, was surrounded by deep river channels on three sides and wetland meadows on the fourth. To the west, expansive mudflats formed the far bank of the river, stretching two miles out to the Bering Sea. During high tides, the mudflats and the river were inundated by salt water, bringing the sea to the edge of our perch, but the camp area itself flooded only infrequently, usually during big autumn storms. Like much of the Y-K Delta, our field site was accessible only by snowmobile in winter and by boat or float plane in summer.

In April 2018, we began setting up our camp earlier than usual to race Mother Nature's accelerating clock. It was a sprint to haul a few tons of gear by snowmobile twenty miles from the nearest village to our field camp before spring conditions made river crossings unsafe. We faced the usual challenges of initiating fieldwork—the "where did we store that last fall?" and the "this motor won't start"—and a few unusual ones as well. During the first weeks of work, constant snowfall forced us to spend more time shoveling and route finding than setting up camp. One day, we were caught in extreme winds and whiteout conditions and had to navigate back to camp by blindly following our GPS track.

By mid-April, the melting tundra and fractured river ice were so impassable that we returned two of our three snowmobiles to town for storage. Then the remaining snowmobile threw a rod through the engine block three miles from camp, a problem that could only be solved by replacing the engine. We had sufficient time to ponder our lack of transportation during an exhausting hike back to camp, postholing through knee-deep snow with every step.

Our field crew was tight-knit from the start. Hayden, the project's PhD student and crew leader, had vastly greater field and life experience than the rest of us. A lifelong waterfowl lover, he had taken a circuitous route to the field of waterfowl ecology. After he completed his undergraduate degree in economics, a debilitating case of Lyme disease forced him to abandon his love of competitive bike racing, so he returned to school to pursue an ecology degree. Despite his atypical academic beginnings, leaders in the field had quickly recognized his potential.

I had met Hayden two summers earlier while working at this same field camp. After fighting the uphill battle of a long-distance relationship for the better part of two years, we had split on amicable terms and managed to maintain a close friendship and productive working relationship. When funding for his PhD project fell through in early 2018 and he could only afford to hire two technicians rather than the usual three, I gladly volunteered to help during the work-intensive first month of the season. I would be working on another project in Alaska for the rest of the summer anyway, so I was glad for the opportunity to help a friend and to spend one more spring on a piece of ground I'd grown to love like my hometown.

Hayden struck gold in hiring Nick and Emily as his technicians—the two became fast friends and worked together flawlessly. Both had earned undergraduate degrees in natural resources and were gaining field experience before considering graduate school. Nick, an experienced waterfowl bander with impressive mechanical skills, and Emily, an extremely knowledgeable birder with an incredible knack for organization, had complementary skill sets. The team would spend the summer searching for and monitoring Pacific black brant nests, the data contributing to multiple long-term waterfowl population studies.

I had completed my master's degree in ecology the previous year, after spending three summers conducting my research at this field site. Though I was the most experienced remote-field-camp occupant, knew the ecosystem as well as anyone, and had an extensive set of practical field and medical skills, I still lacked confidence. During a previous field job, I'd had a tumultuous working relationship with a male coworker. Though I'd been young and inexperienced at the time, I'd worked harder than anyone else to learn the things I didn't know, to sharpen the skills I did have, and to prove that I could be a helpful member of the team. But although I'd earned my supervisor's approval, this colleague had treated me as inept, and humiliated or belittled me at every opportunity.

I appreciated that Hayden never treated anyone as though they were incompetent, instead using their struggles as an opportunity to teach and empower growth. But even under his constructive leadership, I couldn't shake my insecurities. Despite my own internal struggles, though, this group was the most fun and effective field crew I had ever worked with. Emily provided unexpected sarcasm to lighten any situation, Nick provided mischievous hilarity, and Hayden's leadership brought us all together.

On April 28, I watched as a rainbow sheen of diesel began pooling in our Nordic heater—an efficient oil-drip stove system designed for Arctic cabins, yurts, and wall tents. Lighting a tiny scrap of paper towel on fire, I reached my arm into the tubular stove and dropped the scrap into the flammable puddle. The heater roared to life, coughing little puffs of smoke out the chimney as it warmed the WeatherPort, a 12-by-20-foot semipermanent half-cylinder structure that served as our kitchen, gathering place, office, and shelter from the elements. With a metal skeleton and durable covering, it was light enough to be taken down and hauled away every year but also sturdy enough to survive many field seasons in harsh weather.

Emily entered the WeatherPort just as the kettle squealed its morning coffee alert, and Nick came in as the bacon reached optimal crispiness. Hayden was rarely the last to arrive in the morning, so I was surprised when he wasn't right behind Nick.

"What's up with Hayden?" I asked. Hayden and Nick were currently sharing one tent, while Emily and I shared the other.

"His cold is worse today. Don't think he slept much," Nick replied. Hayden had been fighting a head cold for a week, and our arduous trek home from the snowmobile's catastrophic engine failure had worsened it. When Hayden finally joined us, he looked as sick as he felt.

Because funding was limited and satellite phone minutes costly, we'd taken to checking the weather every other day. The forecast, which covered three days, was usually adequate given that the region's weather was fairly constant and predictable, unlike the volatile mountain weather I'd grown up with in Montana. Today being the morning for a weather check, Hayden dialed the number for our area's NOAA coastal and marine forecast after we finished our breakfast.

"More of the same," he said a few minutes later, collapsing the foldable antenna of the clunky satellite phone, a device that looked like a relic of the

1990s. "Wind, snow, overcast. Highs in the mid-thirties today, sustained onshore wind for the next few days."

I geared up for my shift on our daily migration watch, layering on every piece of warm clothing I had, and made the short trek to the observation tower. When I reached the rickety, lichen-covered platform perched fifteen feet above the tundra, I pulled out my Rite in the Rain notebook, donned my binoculars, and fumbled to pull my neck gaiter over my nose. Fully situated, I started my two-hour timer and began scanning the overcast horizon for birds.

Though it should have been the peak of migration, the hours passed without a single flock flying north along the coast. A few groups of cackling geese and Pacific black brant headed inland, carried on the same bitter wind that made my watering eyes fog my binoculars. Little did I know that the birds—or lack thereof—were a harbinger of what was to come.

Lacking birds to document, I distracted myself from the cold by watching my campmates set up our solar and wind renewable energy system. They had nearly finished when I descended the tower. I arrived in time to stabilize the wind turbine's mast as the others equalized its anchor points, tying bowline knots to secure its guy lines. As Emily and I settled into our tent later that night, the wind turbine's roar was a constant reminder of the harsh conditions outside our warm sleeping bags.

By the time I crawled out of my sleeping bag on April 29, the tent's aggressive flapping nearly drowned out the noise of the wind turbine. However, I didn't fully appreciate the strength of the wind until I opened the door to the WeatherPort. The walls billowed as the gust of air rushed in, and the fiberglass door was ripped from my grasp and slammed against the structure's metal frame. I shuffled in, tugging at the door behind me, and it was again ripped from my hand even though I was prepared for the gale this time. With both hands firmly grasping the handle, I adopted an athletic stance and put my full weight behind the struggle. The slamming door finally announced my triumph.

Even with the door closed, cold drafts swept through the usually weather-tight structure. Shielding the burning paper towel scrap with my hand was not enough to keep it from blowing out as I tried unsuccessfully to light the

Nordic heater. I watched in slight amusement as my campmates entered with as much difficulty as I had. By evening, the incessant gale drowned out the crooning of our battery-powered FM radio, so our dinner soundtrack was the clanging of pots and pans and the metallic scrape of the Nordic heater's chimney against the roof.

Because the renewable energy system was making more power than it could store with the sustained winds, we attempted to watch a movie on Hayden's laptop. We crowded around the Nordic heater, propped the laptop on a tote, and chose a movie with maximal action and minimal dialogue. When the movie ended around 10 p.m., Hayden wandered outside, toothbrush in hand, but promptly returned, once again struggling to pull the door closed behind him.

"We have a problem," he said, his face looking ashen. Because Alaska is a wild place, the "problem" could have been a very lost polar bear slinking toward camp or one of our tents pinwheeling across the tundra. We followed him out into the gray haze of a dusky sky and immediately recognized what he was referring to: the river ice was higher than the riverbank rather than three to six feet lower than the bank, where it *should* have been. Through the cracks where the ice flexed, I could see the dark water churning.

Back in the WeatherPort, I looked at the tide chart we kept taped to the door. Glancing between my watch and the chart, I triple-checked the date and time. Tides rise and fall twice each day, with a little more than six hours between each minimum and maximum. Tides also cycle monthly, and tide-generating forces make the highs higher and the lows lower during new and full moons. Much of Alaska's southwestern coast experiences drastic tidal changes—up to twelve feet at our study site. About a week before, the highest high tide of the month had come and gone without fanfare. But tonight, the combination of the high tide, dropping barometric pressure, and sustained onshore winds threatened us with a storm surge.

"We've still got four hours until high tide, probably longer with this wind pushing it in. Our tents are going under for sure. We need to get everything out," I said.

"Put your waders on first," Hayden added. "I think more is going under than just our tents."

I felt a shocked, impossibly calm feeling of awe and disbelief at what we had just seen. Like a line of toy soldiers set to fast-forward, we marched in our waders to our shared tents and began stuffing every nonwaterproof object into dry bags and totes. I slung my belongings over my shoulder and stepped out of the tent with a splash. Spinning around, I paused in horror and saw that seawater was already spilling over the riverbank.

In that moment, we all realized it was going to be more than just an inconvenient evening. It was going to be a nightmare. The water had risen much more quickly than we had expected. It was obvious we would not be able to wait out this storm by battening down the hatches as I had imagined.

The wind began to fling tiny frozen daggers at my eyes, and I struggled to focus. As I scurried toward the WeatherPort in the quickly waning light, dry bag still slung over my shoulder, my wader-booted foot slipped out from under me when I stepped on a swath of now-submerged ice. With a sinking feeling, I imagined my currently dry coat being soaked by icy seawater, but as I flailed, a strong arm caught me and prevented my imminent submersion. Hayden's face was muted terror as he helped me regain my balance.

"We need the skiff. Now!" he said, handing me his own dry bag and pointing me back to the WeatherPort.

I secured our dry bags on the table and rescued a few critical electronics including the satellite phone, the generator, and the computers from the rising water that was now rushing across the floor. Then I dashed back out into the wind and sleet, more careful with my footing this time, to help retrieve the skiff.

Because the river surface was still frozen, we had been storing our eighteen-foot skiff on land with the bow propped up to allow any precipitation to drain out. We had not yet put the outboard motor on since we were weeks away from using it. The skiff had been sitting perpendicular to the wind, and when a strong gust caught the bow, the skiff became a kite and landed upside down. I reached Hayden and Emily just as they bent over our capsized boat. I stuffed my dry gloves inside my jacket and plunged my bare hands into the icy water to help right it.

Once upright, the skiff began taking on water. The boat plug, which had been removed to allow drainage, had been sitting on the transom. In the chaos, the plug had launched into the sea that was now roiling around our shins. As we searched the murky water, blindly feeling around in hopes we'd find the plug close by, I felt a rising sense of panic. Without the skiff, we had no way to save our gear and nowhere to take shelter other than the dilapidated observation tower.

Just then, the first of our equipment totes began floating away. Hayden bolted after one in particular. Grabbing it, he dug to the bottom and emerged triumphantly with two boat plugs, still in factory packaging. He hastily installed a plug, preventing the skiff from taking on more water, just as Nick returned from chasing down the patchwork metal canoe that the rest of us had forgotten about. Within a minute, our fortunes had pivoted from being boatless and stranded, and I was able to tamp down the brief panic I had felt.

Emily and I pulled the skiff toward the WeatherPort, struggling against the surging ocean's furious deluge. I tossed her the pot that had been sitting on the kitchen counter—noodles still stuck to the sides—and she hastily began bailing water out of the skiff. I heaped our personal gear, electronics, the generator, a jerry can of gas, a five-gallon jug of drinking water, and a tote of food in the skiff. A few minutes later, Hayden and Nick returned with the canoe, filled with loose items they had rescued from the surge.

The frigid water had risen to above our knees now. Between the wind and the swift current of water racing inland, Emily and I were struggling to hold onto the skiff. Nick handed out the Mustang Survival flotation coats we usually wore for long-distance boating. We took turns bracing ourselves and holding the bows steady as the others donned the bulky bright-orange jackets.

Then we encircled the boats like a council around a table, unsure what to do next. As much as we didn't want to abandon our camp, we were in a vulnerable location that was receiving the brunt of the storm surge. The nearest high ground was miles away and darkness was settling upon us. We knew we needed to get ourselves out of the water before it topped our waders. We had two options: spend the night in the observation tower and hope that our combined weight and the water's advance didn't topple it, or move inland as far as possible. Hayden pointed inland; less than a half mile away, a large piece of ice rose above a still-frozen stretch of river. We decided unanimously and began our trek toward the ice.

Our traverse was slow. With obsidian water obscuring each step, I slipped twice on frozen ponds. Clutching tightly to the skiff, I managed to arrest my fall each time. Hayden wasn't so lucky. Leading our deliberate charge, he suddenly plunged into an unseen hole and found himself in water up to his chest. He assured us that the water hadn't filled his waders, but the rest of us trod carefully around the spot.

We had all submerged our hands and forearms in the icy water at some point, and our gloves were soaked from gripping the skiff gunwales while ocean water splashed us from below and sleet pelted us from above. As I faced into the wind, its force stole the breath from my lungs. I couldn't tell if the moisture on my face was from the sky, the sea, or the tears the wind had ripped from my eyes. By now, I didn't think we were likely to be swept away by the current, but I knew our fingers, toes, and noses were vulnerable to frostbite, and our bodies' core temperatures were dropping dangerously as the weather worsened.

We trudged for what felt like hours in a semi-wakeful state. Before the black of night fully engulfed us, we finally reached the intact ice we'd been aiming for. After testing it gingerly to ensure it would support our weight, we hauled ourselves up. Though only about twenty feet wide, the strip of ice extended far beyond the beam of my headlamp and seemed to be at least four feet thick.

Satisfied with the ice's stability, our next step was to haul up the skiff and canoe. We formed an assembly line to unload our gear onto the ice and then all dragged up the empty skiff, digging in with our frozen toes for traction.

"Let's use the skiff as a shelter," Hayden suggested, positioning the bow into the wind. I rummaged through one of the salvaged totes, knowing I had seen one labeled "TARPS," and found what I was looking for: tarps and rope. Freezing fingers fumbling with knots, we fashioned a makeshift roof over the skiff, and then piled into our tiny shelter. But four people are not meant to fit comfortably in the hull of a skiff. I could only assume a fetal position, and my legs and feet quickly began to feel numb as I lay on the metal hull with only my soaking-wet floatation coat for insulation.

For the first time in hours, we were out of harm's way, but we were far from out of the woods. "Hayden," I said, nudging him with my foot, "we're safe, but I think it's time to call for help." The full impact of our situation was finally registering for all of us, and I knew Hayden felt an immense sense of responsibility for our safety. Right now, not a soul in the world knew the peril we were facing. With all but a few gallons of our drinking water lost, most of our food supply swept away, and our submerged outboard boat motor likely destroyed, we wouldn't be self-sufficient for long. We needed to let the project supervisor know what was happening.

Hayden hunched across from me, clutching the waterproof satellite phone case, his face expressionless and distant as it often was when he was debugging a complicated piece of statistical code. He was still recovering from the illness that had been plaguing him for days, and I knew he was exhausted.

"Are you okay?" I shouted as a strong gust of wind tore our flapping tarp roof.

I could barely hear Hayden's contemplative reply. "We could have died, Lindsay. What if this storm had hit while we were sleeping?"

I let him play out the many less fortunate fates that could have befallen us, while I tried to think about something other than being cold and dying slowly. I summoned my memory of the humid heat and sweet floral smells that had enveloped my family as we stepped off a plane to visit my aunt and uncle in Hawaii. I tried to recall the feeling of trail dust sticking to my sweaty face on a late July day as I marched up a glacial moraine in Montana. I attempted to conjure the glorious warmth of a natural hot spring on Idaho's Salmon River. Despite my efforts to escape into warm memories, the reality of my shivering body, chattering teeth, and numb fingers persistently intruded on my thoughts. After a while, I realized that I couldn't wiggle my toes. It felt like my neoprene booties had frozen solid.

"I need to move around," I hollered again. "I'm getting too cold just sitting here. Call, okay?" I directed at Hayden and the satellite phone.

"I'm coming, too," Emily said, following me out of a gap in the tarps. Nick emerged right after her.

Nights on the Yukon Delta are the darkest I've ever experienced. This land is a long way from any light pollution, and the sky is usually too cloudy to show a hint of stars or a tendril of the northern lights. It should have been the blackest of nights, but instead, clouds flickered over a luminous waning moon, bathing us in light and then plunging us back into relative darkness. The ice beneath us seemed to glow, and the white streak of an ice wall loomed on the horizon.

The three of us stood with our backs to the wind, jumping up and down, squatting, and shaking our arms like a group of runners on the starting line. Feeling drunk tired, I began to sing without inhibition. "You put youuuuur right foot in. You take your right foot out. You put your right foot innnn. Then you shake it all about!"

Nick and Emily helped me finish "The Hokey Pokey," and we all giggled as we enacted a variety of warmth-generating dances like disco, the lawn mover, and the Macarena. In that moment, I was incredibly thankful to be cheating death on an iceberg with *these* people—people strong enough to laugh and dance the cold away rather than quitting and complaining. As we warmed up, the wind died down a little and we could now hear murmurs of Hayden's satellite phone conversation over the flapping tarp. He emerged from the skiff shelter a while later to find us in a penguin-like huddle.

"Do you guys think we need a Coast Guard helicopter from Kodiak at first light? The other option is a private chopper, but it probably wouldn't get here until the afternoon, maybe 2 or 3 p.m. The worst of the wind is over, according to forecasts, but I have no idea what the extent of the flooding's gonna be."

I thought back to the footage I'd seen from the time-lapse camera we had left on the observation tower two years earlier. Storm surges like this, though unheard of in spring, were quite common in the fall. That year, in September, our campsite had been fully inundated in two consecutive high tide cycles. I expressed my worry that the next tide cycle could be just as dangerous as this one, but we all agreed that we were out of dire straits for the moment, and the expense of a Coast Guard rescue was unnecessary if the chartered helicopter could evacuate us before the evening's high tide.

After Hayden finished coordinating our extraction, Emily suggested that we set up the two tiny backpacking tents she had salvaged so we could try to warm up. She and I wriggled out of our waders and waterproof jackets, crammed into a flimsy-looking one-person tent, and slid into our sleeping bags, thankful for the shared warmth and not caring about the invasion of personal space. I treated my mummy sleeping bag like a sarcophagus, covering my face with the hood to heat the air with my breath. I was still cold and again focused on warm thoughts. Hayden and Nick still hadn't finished shuffling and rearranging themselves in their two-person tent before I succumbed to my exhaustion.

Groggily, I woke to the strangest sound on April 30. *Keeeee-ar. Keeeee-ar.* It took a moment to place the shrill call of an arctic tern. Before I opened my eyes, I wondered if last night's nightmare had indeed been a dream. Bird calls above my tent meant all was normal with the world. But when I peeked out of the vestibule, I saw no sign of a tern. In my hypoglycemic, hyposomnic, and nearly hypothermic state, it took me a while to make sense of the scene outside the tent. Everything inland from our camp was still submerged. The mudflats between our camp and the sea were heaped with chunks of ice—some the size of basketballs, some as big as buses, and some in mounds nearly two stories tall. Luckily, our icy perch hadn't broken apart, nor had we floated farther away from camp.

Miraculously, our camp was somewhat intact as well. The WeatherPort had narrowly escaped destruction by a truck-sized iceberg that had settled a few feet away. One of our tents had survived, and the wind turbine was upright and spinning, though the renewable energy system's electrical box likely was fried. Inside, everything below the nearly five-foot-high waterline was covered in an inch of silty mud, including the rubber boots we had kicked off before donning our waders. Things above the waterline—a high shelf of books and a plate of cookies on the buoyant wooden lab table—remained just as we had left them.

We worked furiously until our scheduled extraction, collecting scattered items—bamboo nest markers, boxed milk, tent stakes, and cots—and then securing the salvaged gear in case of another flood. Snow began falling lightly, and I strained to hear the whirr of helicopter blades, hoping that the quickly descending cloud layer would not impede our rescue.

Ke-ke-ke. I snapped my gaze skyward and seeing the red eyes of a Sabine's gull peering down at the bleak scene, knew I wasn't imagining the sound this time. The ear-piercing rattle of a Sabine's gull call is typically an annoyance, as it often precedes a rain of inescapably foul-smelling, purplish-white fecal material. But the return of life to the tundra infused me with a feeling of hope that I desperately needed. When I finally heard the rhythmic thumping of our ride to safety, I was thankful to have awakened to another morning, and I vowed to never again take for granted the opportunity to be roused by bird calls.

Seven days later, we returned to a landscape still dotted with large chunks of ice and many submerged areas. We spent the next weeks cleaning up the aftermath of the flood's destruction. The many monotonous jobs—some of the most difficult work I've ever done—gave me plenty of time to contrast the unrivaled power of nature with my own vulnerability. Despite feeling small and powerless in the face of the unprecedented storm surge, our team had responded with the strength, courage, communication, and level-headed problem solving needed to survive the perilous night.

In the weeks that followed, I realized for the first time that my unique contributions to the team were deeply valued by my teammates and integral to our collective success. This realization slowly renewed my self-confidence and reinvigorated my passion for ecology, both of which had been waning since the hardships I had faced in graduate school.

Even years later, sitting in a comfortably heated office, I recognize that the lessons I learned in the field—lessons about myself, hard work, and collaboration—still shape how I approach difficult problems and my own fears. And every spring, the far north tugs on my heartstrings. I can't wait to answer that call again.

Cody Deane

Lindsay Gray Carlson is a field and quantitative ecologist and PhD candidate at the University of Saskatchewan. Lindsay grew up camping, hiking, and riding motorcycles with her family in the mountains of northwest Montana. Her outdoor-centric upbringing sparked an early interest in wildlife and wildflowers that has never diminished. She completed her bachelor's degree in biology at Montana Technological University and her master's degree in ecology at Utah State University.

During her early career and before the Starlink era connected even the most far-flung places, she spent more than thirteen months over four summers working at remote field camps in Alaska. Since those formative years, she has pursued opportunities to advance her career across the US and Canada but will always call the rugged peaks and windswept prairies of Montana home. When not working, Lindsay enjoys backpacking, mountain biking, and spending time outdoors with her friends and family.

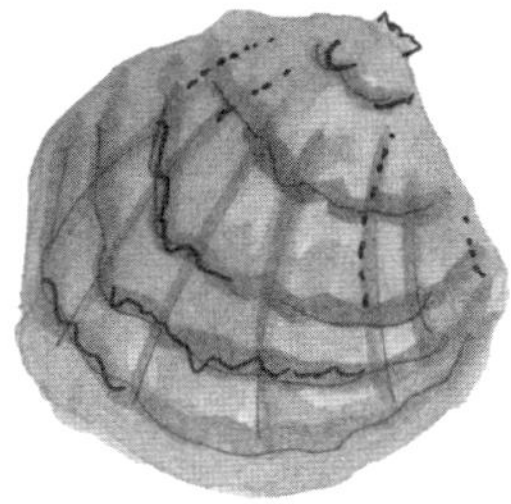

The Importance of Not Being Seen

Emily Imhoff

Early in my career I had a side job working as a technician for a graduate student, Abby, who was studying the distribution, movement, and growth of freshwater mussels. We were working in a large river in eastern Kansas near the border with Missouri, a couple of hours' drive from the university. Abby and I would head out there in Bruiser, the ancient pickup truck assigned to her for use in the field. We'd work by day and stay at a house on a wildlife refuge by night. Since we were traveling, working, eating, and living together, it was good that we got along well and had a lot in common (beyond our shared interest in aquatic invertebrates).

Typically, we would drive Bruiser across the refuge to a put-in at the river, where we would launch our canoe and paddle downstream to Abby's main study site. Then at day's end, we would paddle back upstream against the slow current to the put-in.

One day, however, we had to visit additional sites in a different area, where the river flowed faster, and there was no good spot nearby to launch the canoe. We decided we would put in upstream from these sites, canoe down to them to do our work, then continue downstream to a different takeout spot. However, this necessitated doing a shuttle with a second vehicle. Our plan was to drive Bruiser and a small car to the takeout point, leave the car there, then drive Bruiser and all the equipment up to the put-in. After paddling down to the sites, completing our work, and continuing on to the takeout, we'd drive the car back up to get Bruiser from the put-in, then return to the takeout and load all the gear into the truck. The gear would be unattended for a while, but because of the remoteness of the takeout point, we weren't too worried about anyone stealing anything.

The weather was cool and sunny as we paddled down to the first study site. Abby snorkeled in the shallow parts of the river looking for mussels, identified the species of each one she found, and measured their shells. I recorded the data she collected and glued a number tag onto each mussel so that we could identify individuals and track the mussels' growth if we ever caught them again. I didn't really get to see the animals themselves, since they "clammed up" as soon as Abby pulled them from the water, but it was fascinating to see all the different shapes and color patterns of their shells.

Once we finished at the last site, we packed up and paddled downstream. It was early evening when we arrived at the takeout point, where the car awaited us. However, we soon discovered a problem—the car keys were nowhere to be found. Abby had the keys to Bruiser but not the car keys. To her dismay, she realized they must somehow have been left *in Bruiser*. We were stranded in the middle of nowhere.

We attempted to break into the car, to no avail, and then considered our options. We could try paddling back upstream, but even though we were good paddlers, we quickly dismissed this idea—we didn't have much in the way of lights, and it would be very dark very soon. Not to mention that the river's current in this stretch was fairly fast, and it would take us hours to get back upstream. Calling for help was out, because even my trusty old Nokia phone didn't get service in this isolated place. The final option was to travel on foot to the put-in to get Bruiser. Walking would take a long time, as it was several miles away, and the other downside was stranger danger. We were in the middle of nowhere, yes, but we would be walking on public roads, so we might encounter another human. They would likely offer help, but there was absolutely no way we were getting into anyone's vehicle.

However, it turned out we had a secret weapon—something we had in common but hadn't shared with each other yet. Abby brought it up first. "I used to run cross-country. I can run up there and get Bruiser. I'm sure I could be back in less than an hour." No surprise, I thought, looking at her. Fairly tall, somewhat lanky, but not short on muscle.

I wasn't too keen on sitting around waiting, and splitting up seemed to add another layer of potential chaos to the situation. So I played my own card, which was, in fact, exactly the same as hers: "I used to run cross-country, too. How about if we go together?" She seemed to assess me for a bit, then agreed. We checked the map to confirm our route, triple-checked that we had the keys to Bruiser, and set off running.

By now the sun was at the horizon. We headed up the refuge's two-track dirt road and onto a public road. The gravel crunched under our shoes with each step, keeping time as we engaged in that most primal of human activities—covering ground on foot. We had been paddling much of the day, so our arms may have been tired, but our legs were pretty fresh. It would have been a nice nighttime jog if it weren't for our constant dread of seeing approaching headlights.

Now that we were on an actual road, we needed a plan in case a car *did* come and the driver offered to help/murder us. I had grown up in a similarly isolated area and had avoided being seen by passing cars as a hobby, figuring that not being seen was best, unless I needed help. However, my childhood home was a hilly forest. We were now on a windswept prairie, with few trees or bushes to hide behind or ravines to slink into, but there were occasional ditches and small hills. We decided that if we saw headlights in the distance, we would quickly assess whether we could conceal ourselves before they got close, and if so, do that. If we saw nowhere to hide, we would simply slow to a walk, act casual, and refuse any offered rides.

We figured that if we were going to encounter anyone, it would most likely happen as we got closer to where Bruiser was waiting, since that area was populated. However, being near the small cluster of houses also would give us a better excuse for refusing help—"We're just out taking a walk . . . in the pitch dark . . . for funsies . . . "

Sure enough, we were maybe a half mile outside of the tiny hamlet when we spotted distant headlights coming toward us. There was nowhere to hide, but we were close enough to the houses that we felt pretty good using our "We're just out for a walk" approach. We slowed to a casual walking pace.

Eventually the car approached and pulled up alongside us. The driver, a middle-age guy in a T-shirt, leaned over. "You girls need help? Whatcha doin' out here?"

Abby was ready with her response: "Oh no, we're fine, thanks. We're just taking a walk."

The driver hesitated. "You sure you don't need a ride?"

"Yep, we're good! Thanks, bye!" She waved and turned back up the road, and I followed.

We walked until he was out of sight, then resumed running through the cool night air. We passed the houses quickly and arrived at the clearing by the river where Bruiser waited. It was time for the moment of truth. Abby unlocked Bruiser, rustled around a bit, and produced . . . the car keys! Laughing with relief, we climbed in. Our previously unknown common history had saved the day. Together we fetched the car and our gear, then headed back to the relative safety of the refuge house, where dinner and sleep awaited.

Bob DiStefano

Emily Imhoff was raised by biologist parents in the forests of the Appalachian Plateau, where she enjoyed roaming the hills, wading the streams, and herding chickens. Her appreciation for the varied life-forms found on Earth, particularly in freshwater streams, led her to pursue a bachelor's degree in biology from Oberlin College in Ohio, during which time she completed a research project on crayfish ecology and interned at the Ohio EPA. After working for a Missouri astacologist (crayfish biologist) for a year, she studied crayfish yet again in graduate school, earning a PhD from the University of Leeds in the UK.

Emily has since worked throughout the Midwest for various state agencies, universities, and a natural history museum. She would not be who she is today (and would not have a favorite crayfish—*Cambarus hubbsi*) without the many remarkable mentors who shared their expertise and enthusiasm over the years. She hopes she can be even half as inspiring as they are.

There and Back Again: A Wolf Biologist's Tale

Erin Stahler

"I'll go. I can make it."

Our helicopter circles around to where the wolf is lying in the snow, about a quarter mile into a patch of burned trees. The snow is deep, and, unfortunately, we decided before taking off not to bring our snowshoes.

"Yeah, I can make it," I say again, unconsciously trying to convince myself as much as the other biologists on the crew. "I have a capture kit with a radio collar in my backpack." I'm in Yellowstone National Park embarking on our annual winter effort to radio-collar wolves so we can monitor their population and study how they interact with and influence other species in the ecosystem.

Our original goal had been to collar only one wolf in this pack, but things don't always go as planned. Earlier, our small spotter plane had located the pack we were targeting and had begun circling high overhead to let us know when and where a wolf we'd hit with a tranquilizing dart had succumbed to the drugs. Our helicopter pilot had dropped off our team of biologists nearby to lighten the helicopter's load and increase its maneuverability, and then had begun pursuing the wolves.

The first wolf the helicopter chased was a male known as 1117M that had been radio-collared before, so this wasn't his first rodeo. 1117M immediately recognized the unmistakable *whap whap* of the helicopter and sprinted from the open meadow toward the protective cover of the trees. Before the wolf reached the tree line, my colleague (and husband) Dan fired the gun, sending an immobilization dart zipping toward 1117M. But Dan didn't see the dart hit its target, so the helicopter peeled off and pursued another wolf from the same pack that was still running in the open. Dan aimed again and fired off another shot, successfully darting the second wolf.

After the helicopter swung around to pick up my colleagues and me and head back toward the downed wolf, the airplane's pilot surprised us by announcing that 1117M was down, too, and was lying in the patch of trees. Now, we realize the helicopter can't land amid the trees and the closest drop-off point is a quarter mile away. So we're going to have to get to 1117M through waist-deep snow—not a walk in the park. It's sure to be extremely tiring, and our main challenge will be making it there before 1117M wakes up from the immobilization drug. I've boldly told the others I can make it, but whether I can actually get to the wolf in time, given the difficult conditions, is anyone's guess.

My coworker and friend Elise volunteers to come with me, while the other crew members prepare to radio-collar the second wolf. Elise climbs mountains on her days off, so I have no doubt that she can make the trip with me. A quarter of a mile doesn't seem like a very long way, but the crust on top of the snow is just thin enough that I'm not sure it will support us. We're also carrying a significant amount of gear that adds to our weight and makes wading through snow all the more challenging. After the helicopter drops us off, I take a tentative step, thinking there's a chance the snow's crust will bear my weight, but suddenly I break through, plunging into snow up to my waist. I scramble back onto the snow's surface, but the crust crumbles around me again. I've never experienced quicksand before, but I imagine this is what it must feel like.

We struggle through the snow for a few feet before I try something different. Instead of attempting to stand on the snow's crust, I spread my body weight out by getting on my hands and knees. Yes, I am crawling. We try moving quickly in this posture to distribute our weight and avoid falling through the crust. Like basilisk lizards running across water, we scurry over the snow's surface. Unfortunately, though, we weigh a wee bit more than tropical lizards. So, inevitably, every fifteen feet or so we fall through the crust, face-planting into snow. But to our surprise, this method mostly works.

I yell to Elise, "This must look ridiculous from the airplane!"

I know that my other coworker and friend Kira is circling high above us in the plane, documenting our crazy crawling adventure by photographing us with a long-range camera lens. Elise and I scuttle across the snow as fast as we can, but we can't see where we're going with our helicopter helmets on. Every ten to fifteen feet we stop and sit up on our knees to make sure we're not headed into a tree trunk. We still haven't intersected 1117M's tracks, so I'm worried we're slightly off course. Even a slight deviation wastes time, and every minute that goes by means 1117M is that much closer to waking up and walking away.

I use my radio to talk with Kira in the plane. "Are we still headed in the right direction?"

"Yes, you're about halfway," Kira replies. "If you keep heading in that direction, you'll meet up with his tracks in about twenty to thirty meters."

"Can you tell if he's waking up at all?"

"He's still lying down, but I think his head is up off the snow."

Elise and I start crawling again, and we pick up the pace. I don't care how ludicrous and unprofessional our uncoordinated crawl toward 1117M looks. The chance to radio-collar this wolf is well worth any embarrassment and effort. The data we gather from him will contribute to understanding and conserving his species long after he is gone. We've been following 1117M since he dispersed into this pack, but his first radio collar was chewed off by his pack mates. We have been able to locate him periodically since then, identifying him by his stunning silvery-gray coat in photographs taken from the airplane. (Wolves are born either black or gray; this coat-color transformation to silver is a unique genetic trait passed down through generations in a wolf family, and something we hope to learn more about in the future.)

As we slowly inch our way through the trees that still obscure 1117M, I have plenty of time to think about what might unfold. I reassure myself that all of my past wildlife capture experiences and training have prepared me for the uncertainties of moments like this. Instead of getting flustered or panicked, I am calm, the adrenaline rush of our situation making me think clearly, act decisively, and move purposefully. I have captured wolves, cougars, mule deer, elk, pronghorn, bighorn sheep, bears, and bison. I have experienced wildlife captures that have gone badly, and I have learned from them. I am confident in my abilities, and I thrive in situations that require me to think on my feet.

One of three scenarios might unfold once we reach 1117M: (1) he has metabolized the immobilization drug and is ambulatory, so he runs away from the weirdos crawling toward him, at which point Elise and I do a crawling about-face and head back to the helicopter; (2) he is still completely immobilized, lying on his side with his head down, in which case it will be relatively easy to radio-collar him; (3) he's partially immobilized, sitting or standing up but still not able to walk away—at which point I tackle him.

Despite their reputation, wolves are not the ferociously aggressive creatures fairy tales make them out to be. While bears are known for waking up from immobilization drugs and being aggressive, wolves just want to get away as swiftly as possible. When we handle wolves, we do so as quickly and as quietly as we can to minimize any negative stimuli from the capture. The first sign of a wolf waking up is when it starts to move and lift its head off the ground. Eventually, it stands up and begins to stumble around, which is when all handling by us stops. Wolves do have a bite reflex, and since we don't know when this is regained, it's best to err on the side of caution. I have handled more than a hundred wolves and have never been bitten. However, I also know that when it comes to a wild animal, nothing is ever certain, so regardless of the circumstances, I always act cautiously.

As we approach 1117M, I relay to Elise the possible scenarios we might encounter and review the plan for each. When we're finally close enough to see 1117M, he's looking straight at us. We make eye contact. His yellow eyes pierce right through us. He's sitting up on his haunches, kind of like a dog, watching us approach. I think to myself, *Okay, maybe scenario three.* I stop and draw up an extra dose of the immobilization drug just in case. Then I continue my crawl toward him alone. He sees me but turns his head away so that the closest part of his body to me is his tail. I slowly reach out my hand and give his rump a small but very apparent poke to gauge his reaction. He tenses but doesn't try to walk away. Slowly, smoothly, I press on his shoulder, gently forcing him down to the ground. The extra immobilization drug I drew up is not needed after all.

I then motion to Elise to come and join us. We silently fit him with a radio collar, making sure it's not too tight or too loose. I draw blood from the cephalic vein in his front leg. We take fecal, rectal, and ocular swabs to check for potential viruses. I rub my mitten across the back of his spine, through his soft, silvery fur. He has lost some weight since we captured him last year. I can feel the bumps of his rib bones. I know this will be the last time he's captured; wolves only live an average of four years, and he's six or seven years old. I try to savor every moment. He's the most magnificent wolf I have ever seen.

After Elise and I finish our work—it's taken about twenty minutes—1117M is ready to walk away from us. We gladly let him go, and he stumbles away calmly but awkwardly. As we watch him leave, Elise and I quietly thank him for allowing us to complete our task. The movement and activity data supplied by 1117M's radio collar will help us to better understand how wolves interact with their environment and the important roles they play in the ecosystem.

We crawl back to the helicopter and, exhausted, manage to clamber in. In the past eleven years, that was the farthest I have had to go to collar a wolf; meanwhile, our crewmates processed their wolf a few feet from the helicopter. During the twenty-five-minute flight back to our home base, Elise and I both breathe heavily from the exertion of our lengthy crawl even as we savor our accomplishment. We made a half-mile round-trip through the snow on all fours and looked a half-awake wolf in the eyes while we radio-collared him. It's an experience I'll never forget. But I make a promise to myself to always bring snowshoes in the future—even if I'm sure we won't need them.

Jeremy SunderRaj

Erin Stahler was born and raised in Minnesota and received a bachelor's in biology from St. Olaf College. Her passion for studying wildlife began in Ely, Minnesota, as a volunteer tracking lynx and radio-collared wolves in the Superior National Forest. After mastering the art of radio telemetry, she volunteered on a study tracking newborn elk calves in Yellowstone National Park, and permanently traded Minnesota's thick forests for the open valleys and mountains of Yellowstone.

As a full-time biological science technician for the Yellowstone Wolf Project, Erin helps manage day-to-day operations for monitoring and researching wolves, cougars, and elk. She also serves as a resource helicopter manager and drug immobilization specialist for the park's wildlife capture and radio-collaring operations. Her favorite hobby is looking for antler sheds with her dog outside the park. Erin lives with her husband, Dan, and their son, Rocco, in Gardiner, Montana.

Thanks, Gilles

Émilie Desjardins

Every summer, I travel to the High Arctic to work near a place called Alert, at the northern tip of Canada. Located on Ellesmere Island, Alert is the northernmost permanently inhabited community on Earth. It has hosted military and scientific personnel since the 1950s and can only be reached by air. Because so little is known about the local wildlife and its habitats, I am surveying and mapping the plant communities and recording wildlife sightings to identify critical habitats for Alert's mammals and birds. By identifying local biodiversity hot spots and sensitive areas, I can recommend ways for Canada's Department of National Defence to conserve lands for federally protected species such as the red knot, Peary caribou, and polar bear.

Shaped by the retreat of glaciers, the polar desert where I work is treeless and from a distance seems like an expansive, flat plain. In reality, the terrain is strewn with rocks, intersected by deep canyons carved by glacial rivers, and bounded by snowcapped mountains. And being a frost-molded landscape, even the seemingly flat ground is, in fact, very rough and rugged. Navigating this area, particularly when driving an all-terrain vehicle (ATV), requires a great deal of focus. The risks posed by sharp or unstable rocks, the temptation to look around for interesting wildlife, and the need to be alert for large predators mean that constant vigilance is needed. Traveling safely through areas where polar bears and arctic wolves roam is the main reason to have an ATV in the field.

Each day, I head out to survey the vegetation at various locations and to gather as many wildlife sightings as possible along established transects or in randomly selected plots. As I drive by the shore, I see ringed seals resting on sea ice and arctic terns plunge-diving at the surface of the open water. Sea ducks emit loud nasal sounds while splash-bathing energetically. Long-tailed jaegers fly overhead, making yelping cries that sound almost like puppies. Hunkered down on their nests, well-camouflaged shorebirds flush as I pass by and perform broken-wing displays to distract me from their precious eggs. I glimpse tiny lemmings emerging from their burrows and scanning the sky for danger before attempting any movement, while white, fluffy arctic hares eat every flowering poppy and saxifrage in sight. From time to time, ermines dressed in their brown-and-white summer coats leap curiously around my ATV. More cautious, majestic arctic wolves and slender arctic foxes pause to watch me from safe distances. Sometimes I encounter herds of large ungulates: shaggy muskoxen lying down in the grass or pale, elegant Peary caribou wandering single file, grazing nonchalantly. At the peak of summer, colorful flowers brighten the grayish, bare soil, adding to the unexpected abundance of life flourishing in this harsh environment. But while the High Arctic is incredibly beautiful, its cold temperatures, snowstorms, fog, rain, and powerful winds often make work especially challenging.

One day at the end of August, winds blasted over my survey area at a constant thirty-seven miles (sixty km) per hour. Heavily bundled, head down, eyes watering, I leaned into the powerful wind as I drove my ATV between different plant study sites. At this time of year, the clock was ticking fast. Snow would soon blanket the landscape, putting an end to my field season. Revisiting vegetation survey sites while I still could, I was working my way through my study area to identify some hard-to-differentiate whitlow grasses, members of the mustard family. Stopping at one site, I recorded some key characteristics on a data sheet filled with information I had collected over the course of four days. I was almost finished with my work.

As I went to put the sheet back into my pocket, the wind suddenly grabbed it and blew it out of my hands. Caught up in the strong wind, the sheet took off and soon was barely visible in the distance. I realized immediately that there was no way I could run fast enough to catch it. Jumping on my ATV,

I punched the throttle and sped off as fast as I could in pursuit as I watched the sheet—and days of hard work—disappear over a hill.

Cresting the hill, I spotted a huge male muskox below me, sheltering from the wind close to a large rock. My data sheet tumbled toward him on a collision course with the behemoth. *This is my chance*, I thought to myself. *The paper is going to hang up on him and give me time to get ahead of it.* In spite of the terrain, I drove as fast as I could, watching in astonishment as my fleeing data sheet snagged on the muskox's leg. This gave me just enough time to circle behind him and get into position to intercept it.

Jumping off my ATV, I ran toward the muskox. As a gust of wind blew the sheet off his leg and into the air, I reached out and snatched it just in time. A miraculous recovery! Surprised by my hasty movements, the muskox jumped and squared around to face me. He had not realized how close I was to him—and neither had I. By the time I retrieved my data sheet, I was close enough to reach out and touch him. He loomed before me, blocking the wind and giving me a whole new perspective on his sheer size. For a moment neither of us moved.

Then, keeping my eyes on him, I backed off a few yards and sat down on a rock. We looked at each other for a while. I contemplated the scars on his face, his dark brown eyes, and his broken horn. He had clearly been in some rough fights. I took a picture to remember the moment. Then I returned to my ATV and drove off.

I named my muskox savior Gilles—a popular old-fashioned French name in my native Quebec—and later thanked him in my thesis acknowledgments for saving my data sheet that day.

Courtesy of Émilie Desjardins

Wild animals and vast natural landscapes sparked wonder in Émilie Desjardins from a young age. Her most vivid early memories include hiking with millipedes on her shoulders through the forests of southern Quebec, where she grew up, and spending a quiet day with a pelican floating beside her in the Galápagos. Her desire to protect these beings grew into a deep conviction. After earning a bachelor's in biology from the Université du Québec à Montréal, she began a master's project that evolved into a PhD at the Université du Québec à Rimouski, where she developed a biodiversity management plan for a military station in the Canadian High Arctic.

As soon as she set foot in the Arctic, Émilie fell in love with its immense, windswept landscapes, resilient and curious wildlife, and the rewarding physical and mental challenges of working there. She has returned every summer for nearly a decade. Recently, her postdoctoral research in Arctic conservation biology has focused on protecting Canada's northernmost ecosystems, threatened by climate change and human activity. Her initial sense of awe still guides her work, alongside the conviction that Earth's inhabitants deserve and must receive our protection.

Independence Day in Guatemala

Sophie A. H. Osborn

La Paloma (The Dove), as she was known, had seen better days. Decades ago, perhaps. She was our low-slung, two-wheel-drive, no-extra-cab, rusty-white pickup truck and a critical member of our parrot conservation team. She showed no signs of ever having had luxuries like seat belts, and her headlights had an alarming habit of unexpectedly dimming or dying—a disquieting quirk that sometimes aroused suspicion in anyone who watched her pass by at night. Indeed, this unpredictable behavior had once earned Caín, a senior member of our field crew, a nearly fatal encounter with the heavily armed Guatemalan secret police. Seeing the suspicious approach of his flashing and dying headlights one night, they had forced him to pull over and subjected him to questioning. By a fluke, Caín had recognized one of his interrogators, a long-ago acquaintance from his childhood village.

"*¡Amigo!*" Caín had exclaimed, boldly pushing aside the man's assault rifle and grabbing his free hand in warm greeting. That chance connection had saved Caín's life, and he'd explained La Paloma's foibles to a group of hardened men who had gone from presumptively accusatory to laughingly bemused.

Despite her flaws, we depended on La Paloma to pick up members of our field crew from their scattered homes before daybreak, ferry them to the far-flung parrot nests we monitored during the day, and then return them home after work—singing "La Bamba" in the back of the truck with full-volume exuberance.

I was in southern Guatemala for my second field season, studying and trying to protect the yellow-naped amazon—a large, beautiful parrot the color of new spring leaves, with a splash of yellow on its neck and brilliant red and indigo in its wings. We were investigating the wild habits of this popular cage bird, which was heavily poached to satisfy the demand for pet parrots in the US and Central America. Nestlings were often stolen from their nests and smuggled across borders. The few that survived these ordeals went to owners who rarely knew of their pet's nefarious origins.

Each day, members of our part-Guatemalan, part-American field crew spread out over several large cattle ranches and a remnant patch of tropical dry forest to monitor parrot nests—natural cavities in tree trunks—and record nesting behavior. We documented the number of visits adult parrots made to their nest to feed their nestlings and the amount of time females spent in the cavity with their chicks. But in reality, our observation duties were as much about guarding nests from poachers as they were about collecting scientific data. On one occasion, I watched in horror from a nearby blind as a poacher started to climb the nest tree I was monitoring. I scared him off by blowing a safety whistle.

Despite our best efforts, nearly all of our nests were eventually robbed of their nestlings. Our presence near the nests usually dissuaded poachers during the day, but desperately poor sugarcane workers climbed the trees at night, braving nocturnal dangers to supplement their pitiful incomes. And our field crew navigated the heartbreak of watching distressed parrot pairs fruitlessly searching for their young over the following days, calling softly as they repeatedly went into and out of their nest cavity.

While we nest watchers guarded nests and recorded the parrots' activities, a climbing team, garbed in protective beekeeping suits, ascended each nest tree—braving attacks by fearsome cavity-dwelling killer bees—to measure the nestlings' growth and fit several of them with metal radio collars so we could track them if they fledged.

As the sun rose over a parched shrubland where cattle eked out a meager existence, I drove along a dirt track with the two Guatemalan women who worked on the project, delivering them to their assigned parrot nests. I dropped Mirna off first and watched her walk into the bush, a baseball cap and long-sleeve shirt shielding her from the sun, her waist-long braid unmoving as she strode toward a distant line of trees. Ferociously conscientious, Mirna recorded the parrots' activities and relayed her findings with a no-nonsense demeanor that made the warmth of her rare smiles all the more striking. Her participation in a project that would usually have been the purview of men in Guatemala was a rarity, encouraged by the two American women—a biologist and a veterinarian—who co-led the parrot project and supported local women with an all-female soccer league and work opportunities.

Smiling over at Tehlma, I depressed La Paloma's gas pedal, and we headed to our next nest site. Though the two women were opposites, I treasured Tehlma's friendship as much as I did Mirna's. Softer, shyer, Tehlma was a dreamer who spoke in quiet tones and drew flowers in the margins of her field notebook. Prone to whimsy, she'd had a small silver star cemented to one of her front teeth. Both women were devoted to the parrots, relishing the opportunity to spend time observing them.

As we bumped along the rough road, Tehlma and I chatted in Spanish, pausing periodically while I searched for unknown words and laughing together at my occasional grammar mistakes. We talked about our families and our love of wildlife as we headed toward a towering volcano that loomed in the distance and glowed at night with erupting lava. An elegant scissor-tailed flycatcher called from a nearby bush and then flushed as we passed, flashing its pink underwings and elongated black-and-white tail.

Suddenly, our happy chatter was disrupted by a dull *bang*, followed by a thunderous rumbling that made La Paloma's vulnerable frame shudder. Holding tight to the wheel, I hit the brakes and brought us to an abrupt standstill. Tehlma and I looked at each other in wide-eyed alarm.

"*¡Ay de mí*!" I exclaimed. "That did *not* sound good!"

"*¡Ay, Dios mío*!" Tehlma echoed. "What's happened? I think La Paloma just died."

We sat for a moment in silence before Tehlma continued, "What are we going to do? Where are Marco and Caín? How long do you think it will take them to realize we need help?"

Knowing it would be a long time before one of the kind and capable Guatemalan men on our field crew came to our rescue, I said, "We'll be fine. Let's take a look at what's going on."

Neither of us was mechanically inclined, but I jumped out of the truck, feigning a confidence I didn't feel, and walked around to the back of the pickup. Tehlma was beside me in seconds. Wordlessly we stared at La Paloma's undercarriage, which appeared to be lying on the ground. But as I crouched to take a closer look, I realized it was only the truck's decrepit exhaust pipe and muffler that lay in the dust under the vehicle. Apparently, the bracket that attached the muffler to the underside of the truck had broken. The terrible sound we'd heard was the muffler and exhaust pipe hitting the ground and then scraping and bumping along as we drove.

"It's only the muffler," I exclaimed in relief. "I think if we can just tie it back up, we'll be fine." Tehlma looked at me dubiously, then peered into the empty truck bed.

"We don't have anything to tie it up with."

"There must be something," I said as I rifled behind the truck's bench seat. I found nothing. Returning to the back of the truck, I glanced over at Tehlma. I had never witnessed someone literally wringing her hands in distress, but now I watched as Tehlma clasped her hands, fingers interlocked, and rubbed her palms together.

With a furrowed brow, she again looked at the muffler, then repeated a little plaintively, "What are we going to do?"

I suspected that she was mainly concerned with the immediate dilemma of having broken down in an isolated spot, leaving us unable to do our work or return to our small field station. But fears for our safety were not unwarranted. A female sugarcane worker had allegedly been beheaded during the night on a nearby ranch months earlier. Guerrillas and military personnel periodically burst into conflict nearby, while keeping not-so-surreptitious eyes on our activities. Local police sometimes stalled us with roadblocks on the highway until we succumbed to their demands for bribes, which I learned—after a frightening solo encounter with officers who threatened to impound La Paloma—usually amounted to giving them a few cold sodas. In addition, bandits were omnipresent enough at night that our Guatemalan crewmates expressed relief when the secret police were in the area "removing" anyone they believed was acting suspiciously. Despite these mostly unseen threats, the lazy rhythms of the ranches where we worked, the rarity of running into anyone while we observed our parrots, and the good humor and affection bestowed on me by our Guatemalan crew members made me feel safe.

"Don't worry," I smiled encouragingly as my mind raced to find a solution when none seemed in the offing. "There must be something we can use to tie up the muffler." Returning to the front seat, I rifled through my pack.

"Tehlma!" I exclaimed moments later. "I found something." Pulling out my compass, I triumphantly detached a slender red cord that allowed me to wear the device around my neck.

Tehlma laughed, eyeing the cord skeptically. "It doesn't look very strong."

"I know," I agreed. "But it might work. The only other thing I can think of is twisting strands of grass together to make a rope, but I'm afraid the grass might get hot and catch fire."

Lying on my back, I maneuvered myself under La Paloma and searched for somewhere to attach the cord. After anchoring it to the undercarriage, I asked Tehlma to lift up the muffler, then wrapped the thin cord around the attached exhaust pipe and pulled the muffler back into its usual position. After tying several knots, I wriggled out from under the vehicle, smiling triumphantly. Extending her hand, Tehlma helped me to my feet, her smile mirroring mine.

"It worked!" she exclaimed. We gave each other a high five.

"See," I laughed in relief, "we can do anything! We don't need help from the guys."

"We can do anything!" Tehlma repeated happily, flashing her silver star again.

"Well, maybe we better make sure it holds up while we're driving," I cautioned wryly.

Climbing back into the vehicle, I restarted La Paloma and headed slowly toward the nest Tehlma would be observing. There was no thunderous rumbling, no disheartening crash—just the usual creaks and rattles and rumbles of La Paloma's familiar travel song.

That evening, when our field crew gathered as usual in our small dining area to chat about the day's events before half the team headed to their homes and the rest of us ate dinner and retired to our cots, Tehlma regaled everyone with the tale of the great muffler debacle. Grinning, I listened with rapt attention as usually timid Tehlma elicited laughter while describing the teeny, tiny compass cord and what we had achieved with it. Her pride in our resourcefulness permeated the festive atmosphere. As she finished her story and her animated hands settled by her sides, Tehlma and I locked eyes and beamed at each other. Our incandescent smiles bore an unmistakable message: it was Independence Day in Guatemala. And that confident, star-toothed smile was an image of empowerment I'd carry with me for the rest of my life.

Courtesy of Sophie Osborn

Sophie Osborn grew up in Switzerland and Vermont, surrounded by pets. It wasn't until after she earned a bachelor's in international relations and French literature from the University of Pennsylvania that she discovered wildlife biology. Changing course, she studied wildlife in Africa and the US. Her first summer of work as a wildlife biologist, reintroducing captive-reared peregrine falcons to the wild in Wyoming, inspired her to dedicate her career to helping birds.

Sophie subsequently worked on projects with a variety of raptors and songbirds—including the American dipper, which she studied for her master's degree at the University of Montana. She also worked with parrots in Guatemala, ducks in Argentina, and cloud-forest eagles in Peru. Her book *Feather Trails: A Journey of Discovery Among Endangered Birds* chronicles her experiences reintroducing peregrines, Hawaiian crows, and California condors to the wild. She also writes the *Words for Birds* blog. Sophie lives in Montana with her two beloved border collies and an irrepressible orange cat. She loves hiking, running, reading, birding, traveling, and creating a refuge for wildlife around her home.

Awe and Enigma

Stalker

Susan B. Adams

Independent, hardworking, goal driven—that was me. In graduate school, I hadn't yet learned the difference between working endlessly and working efficiently. So I just worked, putting in long days in the field over two intense summers. When my field assistant headed home for weekends, I often slogged on by myself. I was lucky to be doing my summer graduate research as an employee of a federal agency, but at one point my supervisor insisted that I not work alone in the field anymore.

"Okay," I agreed. *Sure, whatever*, I thought to myself, knowing full well I would do whatever it took to complete my research.

It was nearly the end of my first summer of fieldwork for my master's project assessing habitats used by the small bull trout that spent their lives in cold mountain streams. The species was declining, and my results would help agencies better conserve and restore their habitats. My assistant had returned to the university a couple of weeks earlier when the semester started, but I still had two days of work left to complete my planned surveys of the fish and their streams. Naturally, I did not hesitate to go out alone. I drove two hours through the mountains of central Idaho to one of my study sites, set up my tent in a little unofficial campsite, drove about a mile to the last stream section I needed to sample, and got to work.

I spent the remainder of the morning measuring characteristics of Dewey Creek, a small, cold, spring-fed stream. One set of measurements involved recording water velocities at multiple points across the creek. Being an underfunded graduate student, I lacked a fancy electronic velocity meter and instead had to use an old-fashioned pygmy meter. This was a temperamental device attached to a four-foot rod that I held vertically, with the bottom end resting on the streambed. The instrument's small fins spun in the water like a horizontal windmill, with each rotation creating a little *click* sound.

Wearing a pair of headphones, I counted the number of clicks the meter made in a set amount of time. The faster the water was flowing, the more clicks I heard. Most. Tedious. Work. Ever. But the scenery was pleasant: a variety of conifers rose from dry hillsides; dense patches of fir, spruce, willow, and alder dotted the stream banks; and the water gurgled along peacefully. My head wasn't in the game, though. I was tired, looking forward to wrapping up the field season and not pulling on a pair of waders again for a very long time. My thoughts were more on the university classes I was missing than on my surroundings. The headphones added to my oblivion.

About midday, when the air neared its warmest, I switched over from assessing stream habitat to finishing my snorkel surveys for bull trout. Leaving my habitat-measuring gear by the creek, I hiked back up the steep slope to the truck and donned a layer of warm clothing. Then I pulled on my red, white, and blue dry suit that someone had dubbed my Captain America suit. The dry suit's tight rubber gaskets kept water from seeping in but squeezed my ankles, wrists, and throat uncomfortably. The knee pads I wore over the dry suit made it difficult to bend my legs.

After waddling along in my bulky attire for about the length of a football field, I veered off the dirt road and made my way clumsily down the slope and through a streamside thicket of alders. I plodded over to the flagging I had tied to a branch the week before, indicating where I had stopped snorkeling. There, at the downstream end of the section I still needed to snorkel, I prepared for one of my last chilly bull trout searches of the season.

To minimize the ice cream headache I inevitably experienced when submerging my head in the frigid stream, I pulled on a thick, snug neoprene hood, tearing out a few long hairs in the process. The neoprene gloves I put on next made everything challenging but usually kept my fingers warm enough to hold a pencil and tally fish on a small plastic slate. Finally, I put on the mask and snorkel that allowed me to search for the small adult and juvenile bull trout that hid in the stream's secret places.

The thought of snorkeling might conjure images of someone floating along like a duck, peacefully looking for fish in serene crystalline waters. But the knee pads I wore hint at the much different reality. Snorkeling this tiny stream involved crawling upstream on my hands and knees over the rocky streambed and tilting my head sideways to slowly peek into any pocket of water deep enough for me to submerge one eye. Where the water was deeper, I sometimes had to submerge my head—icy water filling my ears—to see into the calm water underneath the bubbly white turbulence nearer the surface. The few "large" pools, perhaps the size of three kiddie pools, provided a real treat—there I could rest comfortably at the pool's outlet with both eyes in the water and peer secretively into that hidden world. Sometimes I would just lie in one spot, unmoving, for several minutes, waiting for the fish to leave their dark sanctuaries and resume feeding so I could count them.

The burbling stream and my thick hood blocked many terrestrial sounds, and while wearing my mask, I couldn't see clearly when I lifted my face from the water. The stream realm became my entire world, my focus—immersing myself in it was truly one of my greatest loves. Sometimes when I was stealthy and patient enough, the fish seemed to forget I hadn't always been there. In those magical moments, they would swim to where the water rushed by my boulder-like head and station themselves in the fast water to feed, tail fins gently tapping against my cheek. That summer, my exhausted body was lulled into slumber on many nights by visions of those little bull trout looking at me or patting my face with their caudal fins. Decades later, I sometimes still see them.

As I crawled and snorkeled my way upstream on this penultimate day of surveying, I noticed an unusual amount of gunk floating down the creek. Although the stream could be clouded by fine suspended silt after a rain, we were in the midst of a long dry spell. I usually watched fish through clear water. Nevertheless, conifer needles, leaf fragments, and small clouds of silt drifted down the stream toward me. Something seemed to be disturbing the substrate upstream from where I was snorkeling. Curious, I lifted my face out of the water several times and eventually removed my mask for a better look, convinced I'd see a cow standing upstream or crossing the creek since this was free-range cattle-grazing country. But there were no cows, no deer—nothing in the water upstream. At least, nothing that I could see.

After about twenty minutes of snorkeling, I came to a deeply recessed undercut bank—a favorite habitat of bull trout—and realized with disgust that I had left my waterproof flashlight in the truck. Even at noon on a sunny day, an underwater light was essential for finding fish in the shadowy recesses of stream banks and logjams.

I trudged back up the hill to the road, annoyed at my forgetfulness. It had been a long summer. As I plodded up the road to my truck, I contemplated the three-hour drive back to the university, the statistics assignments that awaited me, and the future coffee dates I would have with friends. The neoprene hood muffled sounds, facilitating my wanderings inside my own head.

After retrieving my hidden truck key, unlocking the door, and grabbing my flashlight, I turned around to replace the key—and froze.

Standing in the road not fifteen yards from me was a mountain lion. A *big* mountain lion. A big, gorgeous mountain lion like the one I'd dreamed of seeing since I was a little girl. Waist high, tawny, intense—and undoubtedly much faster than me! I was ecstatic and yet thoroughly relieved I hadn't seen him until I was standing with one hand on my truck. We locked eyes for a moment, and then he slipped silently down the hill toward the stream, hiding behind some bushes and a large tree about twenty yards away.

He peered from behind the tree trunk, watching me and flicking just the tip of his tail like an overgrown house cat. I had not spent much time with cats, but that intense tail twitching seemed to be the type of behavior that comes before a big pounce. Happily, I was out of pouncing range (or so I hoped), and having my truck between the lion and me added to my sense of security.

When I leaned sideways so that the tree hid me from his view, he leaned out to keep me in sight. When I started overheating and removed my hood, he sniffed the air. When I pulled off the top of my dry suit, he sniffed again. Perhaps I was naive, but I thought he seemed more curious than threatening. After about five minutes, I walked very slowly around to the lion's side of the truck to get my camera so I could get a closer look at him through my telephoto lens. As I opened the door, a truck drove by. I took a quick look over my shoulder at the passing vehicle, and when I turned back, the lion was gone. I was crushed, not wanting our encounter to end. I desperately wanted more time to observe this beautiful, elusive creature.

For another five or ten minutes I stood by my truck, hoping for a glimpse of him peeking out from behind a shrub or tree. I was just about to give up when I spotted him again. In the time it had taken me to glance at the passing truck, he had moved into a thicket of willow and fir trees—about five yards *closer.*

Realizing how long it had taken me to find him again, even though I knew he was around, freaked me out. Seeing him through my camera's telephoto lens freaked me out even more. His chest was wet. It was late summer in dry country. The only place he could have gotten wet was in the very stream in which I'd spent the previous twenty minutes lying face down.

I immediately recalled the detritus I had noticed floating down the stream earlier. Had the lion been watching me as I lay face down in the stream, absorbed in the underwater world and heedless of my terrestrial surroundings? Was he focused on me, trying intently to figure out what the heck I was . . . and whether I was edible? No doubt I looked strange to him, and encased as I was in a dry suit and hood, I probably didn't smell like other humans he may have encountered previously. Maybe he'd been watching me all morning while I counted clicks of the pygmy meter and ruminated about the classes I was missing. And he'd certainly followed me—maybe even stalked me?—from the stream all the way back to my truck. Most unsettling was that I'd never had even the slightest sensation of being watched.

When I thought about the lion silently padding behind me as I walked to my truck, I laughed at what the occupants of the passing pickup might have thought had they rounded the curve minutes earlier and seen a young woman in a patriotic Captain America dry suit walking along a dirt road in the middle of nowhere followed by an adult mountain lion.

Eventually, another vehicle drove by. Again, I turned to glance at it, and again, when I looked back, the lion was gone. I never saw him again. But I was certain that he was still watching me.

Now I had to decide. From where I was standing at my truck, I could see my habitat sampling gear down by the creek where I'd left it earlier. Should I retrieve it? On the one hand, the lion had already had plenty of opportunities to attack me, and yet, here I was, all in one piece. On the other hand, my supervisor had declared that it was too dangerous for me to work alone, and

I had no interest in proving him right. I imagined the headlines: "University of Idaho Graduate Student Eaten by Mountain Lion; Family and Friends Devastated."

I left the sampling gear, drove up the road, packed up my camp, and drove the two hours home to McCall, Idaho. As I made my way back to town, I couldn't stop thinking about the lion's stealth and curiosity. Mountain lion sightings were rare in those days, and I felt like I'd received a tremendous gift—a grand finale to end the summer.

The next day, I returned to the field site with my boyfriend's dog. Tally stood guard as I gathered my gear and finished that work in record time. I avoided the headlines, but never have I looked over my shoulder so often!

Author's note: I later learned that two men marking timber the previous summer just over the ridge from where I'd been snorkeling had also encountered a mountain lion. The lion had circled them, moving continuously closer, and the men had ended up with their backs against a tree before finally making a break for their truck. We figured it was probably the same lion—likely a young adult male that was either unusually curious or hungry.

Courtesy of Susan Adams

As a wild child in Northern California, Susan (Susie) Adams was enthralled with nature, an inclination nurtured by her mom, who started a Ranger Rick Club for neighborhood kids. Susie grew up riding her pony and bike on neighborhood trails, and swimming and fishing in local ponds. Wanting to experience *real* winters, she attended Carleton College in Minnesota, earning a bachelor of arts in biology in 1986. After working as a ski instructor, helitack firefighter, and fisheries technician in Idaho, she returned to school for a master's in fisheries management and then earned a PhD in ecology from the University of Montana.

With years of trout and salmon research under her wading belt, she moved to the only US state with no salmonids—Mississippi. Working as a research aquatic ecologist and team leader with the US Forest Service in Oxford, Mississippi, she has gradually shifted her research focus from fish to crayfish. Susie is happiest in, on, or near water, and her favorite possession is her aging teal-green kayak. She loves adventuring with her friends and two grown kids, reading, and empowering women to gain confidence and experience joy in the outdoors.

A Whale of a Tale

Deanna (Dede) H. Olson

"Humpback whale!" I heard someone exclaim. We all rushed over to the cliff's edge to peer down at the sheltered inlet two hundred feet below. There it was, swimming a lazy circuit of the cove where our boat was anchored. Our boat captain, Felipe, who had remained on board, was watching it as well.

It was our second day on San Pedro Mártir Island, Mexico, in the Sea of Cortez, also known as the Gulf of California. I had just turned twenty years old and was helping a doctoral student with his research over spring break, along with five other undergraduate students from the University of California, San Diego. The San Pedro side-blotched lizard, which we were studying that week, occurs only on its namesake island. The most remote island in the Sea of Cortez, San Pedro Mártir is tiny—about one square mile in area—and is located thirty-two miles east of the Baja California peninsula and thirty-three miles west of the Mexican state of Sonora.

The island's ecosystem is unique. As we approached by boat the previous day, I was struck by the steep stark-white cliffs framed by the deep-blue sea below and the sky-blue expanse above, with the occasional dark-green columns of cardon cactus piercing the skyline. The white that colored the cliffs was guano (bird excrement), deposited so thickly over the ages by the constant eddy of seabirds flying between water and land that it was harvested for fertilizer in the early 1900s. The island's ever-present gulls and pelicans were familiar to me, as were the California sea lions that dotted the shoreline rocks, so my eyes more often focused on the island's novel and colorful sights:

the occasional flash of a male magnificent frigatebird's red throat pouch or the pale-blue webbed feet of the blue-footed boobies. Males of both species use their colorful parts to attract mates—frigatebirds by inflating their striking throat pouch with air and boobies by lifting their vibrant feet in courtship dances. In 2002, San Pedro Mártir Island was designated a UNESCO Biosphere Reserve. It is among the most important bird sanctuaries in the Sea of Cortez.

After watching the humpback whale circling in the cove, we caught intermittent glimpses of it throughout the day, as well as later that evening from the boat, where we retreated to eat and sleep. Because of its charismatic presence, it quickly became the unofficial mascot of our trip. We saw it again on our third morning as we trekked up a trail that ascended the cliff and led to the adjacent plateau.

Through our field observations, we were trying to understand the island lizards' social relationships and interactions to compare them to those of mainland lizards, which had more predators and competitors. Individual lizards seemed to have established territories around the blue-footed booby nests, which were strewn along the rocky cliffs and the adjacent plateau. The booby nestlings had hatched, and as parents regurgitated fish to feed them, the lizards darted in to gobble up both the messy fish scraps around the nests and some of the attendant flies that clouded around the nestlings' leftovers.

Lizards near nests were aggressive toward apparent interloper lizards and indicated their dominance via behaviors such as head bobbing and chasing. Interactions among the larger male lizards were particularly dramatic, with frequent chases between territory holders and interlopers. Our observations suggested that male lizard territories were larger than those of females and possibly encompassed several female territories. Perhaps males helped protect females in their territories, restricted other males from mating with them, and provisioned females with food. To examine these relationships, we were arrayed across the plateau in adjoining zones along the cliff's edge, and each of us charted lizard movements and interactions within our area. We each had at least one booby nest in our zone so we could see how nests affected lizard activities. Because the lizards were extremely active, we accumulated data very quickly.

Before the San Pedro Mártir study, we had all worked on a study of the closely related common side-blotched lizard in the San Jacinto Mountains of California, so we were well-seasoned "lizardeers." Having that experience made our current job easier since we needed to mark the island lizards' backs with paint so that we could visually identify individuals from afar. Although some researchers preferred to mark the fast-moving lizards with a paintbrush attached to a stick, that approach often resulted in an unrefined splotch of color across the lizard's back, tail, or leg. I preferred to catch them first so that I could mark them carefully while they were in my hand. I reveled in the thrill of noosing the lizards. Holding a long, narrow stick like a fishing pole with a tiny thread noose attached to one end, I would loop the noose around a lizard's neck and then pull back, like lassoing a pint-sized pony. This stealthy trial-and-error pursuit required a steady hand, patience, and persistence. Too often, the awkward stick would accidentally bonk the lizard in the head, or the wavering delicate thread would twist the wrong direction in a gust of wind. Even more irksome, annoying flies would land on us, wrecking our concentration right as we were about to succeed in our lizard rodeos.

Once captured, the larger island lizards were a handful compared to their smaller cousins in California. These island lizards had few natural predators or competitors and were not only large—up to about seven inches long—but also fearless. As we sat quietly recording their behaviors and movements, they would climb up and perch on us, treating us as part of their territory. People made a good lookout post, I think! Or perhaps we helped them get fast food: I would silently cheer on lizards that pursued pesky flies on my legs.

The memorable interruption to see the humpback whale was one in a string of "Look at this!" moments that punctuated our data collection in this wildlife haven. On the third day, we were well into our lizard work when someone shouted: "Where's Felipe going?" Once again, we lined the cliff's edge and looked down. The boat was moving out of the cove into open water. We had hired Felipe's boat out of Guaymas, almost a hundred miles to our southeast on the Sonoran side of the Sea of Cortez. Since he had ferried previous lizard research trips to the island, Felipe had been our go-to guy, a jovial companion who kept us smiling.

"He's going backward!" someone exclaimed. This seemed odd—and disconcerting. We had no way to communicate with either him or the mainland since the only radio was with Felipe on the boat. We all looked at each other, hoping we hadn't been abandoned. Fortunately, Felipe and the boat returned a couple of hours later. After finishing our fieldwork, we clambered down the cliff trail to the cove to find out why he had temporarily stranded us.

Felipe seemed both bemused and a bit put out. As it turned out, the whale had grabbed the boat's anchor and towed the boat and Felipe out to sea. Not wanting to lose his only anchor, Felipe had decided to see what would happen next. The whale kept going and kept towing. As they moved into deeper water, Felipe became concerned that the whale might dive, pulling the boat down and capsizing it. Finally, opting to abandon the anchor, he cut the anchor line and returned to the cove. Felipe's story prompted wild speculations and a lot of jokes among our field crew. We wondered whether the whale was territorial, like our lizards, and annoyed that we were in its domain uninvited. Or perhaps it was a juvenile-delinquent whale with a sense of humor, just having a lark.

Since we now had no anchor, we rigged up a rope, tying the boat to a rock on shore to keep us tethered. The weather was calm, so we hoped this makeshift mooring would hold us safely in the cove for the rest of our stay. After dinner, as it got dark, we laid our sleeping bags in a row on the deck to enjoy the stars. This was the life!

At dawn on day four, as we still slumbered, the boat suddenly heaved beneath us. Tossed across the deck, we tumbled in heaps over one another. Disoriented, we jumped to our feet, as the deck continued to rock and roll beneath us. As we looked around, trying to find out what had set the boat in motion, we saw the whale circling the boat and creating wakes in the water. It seemed to have breached next to the boat, possibly hitting our mooring line and causing the entire boat to go topsy-turvy, rocking bow to stern. I wondered if it had tried to grab our rope to take us away again. As we were recovering from the boat's violent motion, sitting back on our sleeping bags, a gull crash-landed on the railing next to us. Without warning, it spewed its noxious stomach contents of half-digested fish all over us. Double whammy—this trip was heading downhill fast!

We took stock of the whale situation and how much lizard data we had accumulated, and decided it would be safer for us to finish up our work that morning and leave the increasingly inhospitable cove. Years later I learned that boats in far-flung locations have been subjected to similar antics by humpback whales, breaching near boats and, in one instance caught on video, towing a boat away by the anchor line. Even today, there seems to be ample speculation but few facts about what motivates such behavior.

After trudging up the cliff that last morning, I looked down from atop the plateau and saw Felipe moving the boat to moor it safely outside the cove. Meanwhile, the whale patrolled the cove, spouting air out of its blowhole and slapping the water with its tail flukes as if cockily declaring its ownership. A few hours later, we motored out of the cove for the last time, leaving the whale to its solitary domain.

Upon our return home, we had expected to regale our family and friends with amazing stories about lizards, but we left the island with a whale of a tale instead.

Courtesy of Deanna Olson

Deanna (Dede) Olson is an emeritus scientist (retired research ecologist) with the US Forest Service Pacific Northwest Research Station in Corvallis, Oregon. She was born in Michigan, where she bonded with toadlets in the basement window well of her house. When Dede was seven, her family moved to Los Angeles, where she freely roamed local neighborhoods. She earned a bachelor's in biology at the University of California, San Diego, and a PhD in zoology at Oregon State University.

Along with studying the general ecology of a variety of species, Dede studies potential threats to animal populations—especially amphibians and reptiles—including impacts from climate change, disease, and habitat alteration by forest management. She and her husband, Mike, live with their black cat, Harry; labradoodles, Rocky and Rosie; and rescued Russian tortoises, Miss America and El. Besides keeping up with their daughter and son and their spouses, granddaughters, and grand-doodles, Dede gardens, skis, lifts weights, and paints.

Easy Breakfast

Zanethia Barnett

As an African American girl, I never dreamed of scuba diving. The very idea was as alien to me as putting on a spacesuit and flying to the moon. I grew up in a small town on the South Carolina coast, spending a lot of time fishing and playing on the beach, but I never saw a single person in a wet suit or loaded down with an oxygen tank. Scuba diving was something I only saw people doing on television, and those divers were never African American and were rarely female.

So when I had an opportunity to learn how to dive while pursuing a master's degree in ecology, I jumped at the chance. This was my ticket to becoming one of those cool people I'd seen on TV swimming with dolphins and exploring underwater caves. But to become a certified scuba diver, I had to pass a demanding swim test that included swimming underwater for twenty-five yards, swimming four hundred yards within twelve minutes, and treading water without using my hands for two minutes. *How was I ever going to pass?* I hadn't worked out consistently since high school and swam recreationally only a few times a year.

That summer, I practiced and practiced, spending most of my free time swimming. The first time I tried to tread water with no hands, I threw my arms in the air and kicked as hard as I could, but my body slowly started sinking. With a splash, my hands hit the water to prevent me from going under. Panting like a dog after a long run, I grabbed the side of the pool and looked at the timer on my phone. Only fifteen *seconds* had passed! After many more unsuccessful tries, I left the pool exhausted and defeated. That night

I watched YouTube videos and read articles on how to tread water and get into swimming shape. I practiced every day for the entire summer. By the fall semester, I was ready—and I passed that swim test.

After learning in a classroom about the science behind scuba diving, I was finally prepared to try out my new skills. On my first dive, I drifted down into the depths of a spring-fed sinkhole in Florida and watched everything slow down. The hustle and bustle of the world above disappeared, and everything became quiet. Fish and turtles glided past me as if I weren't even there. I was overcome by peace, tranquility, and serenity. I felt as if I had slipped into a new world where my only responsibility was to relax and lose myself in my surroundings. I had never felt anything like this before. I was hooked.

Fast-forward five years. As a freshwater biologist studying how changes in aquatic environments affect the animals that live there, I'd become passionate about crayfish. Alabama, where I was conducting my research, had a hundred crayfish species—the greatest crayfish diversity in the world. As important members of aquatic ecosystems, these lobster look-alikes are eaten by more than two hundred animal species—including people. Found only in fresh water, crayfish are surprisingly diverse and can be brightly colored—vibrant hues of red, blue, or purple—or patterned with spots or stripes. Some south-eastern US crayfish species grow as large as my hand, while others are no bigger than my fingertip. Surprisingly, when I began my research, little was known about crayfish that live in large rivers. Here was a chance for me to combine my two favorite things: diving and crayfish.

Early one morning, two colleagues and I headed out to the Tennessee River in northwestern Alabama to collect crayfish for my research. We were the lone boat on the river that morning. As we drifted downstream, I scouted for large boulders lurking under the river's surface. Crayfish often shelter under large rocks to avoid being eaten by predatory fish. If I was strong enough, I was going to roll some boulders over to expose hidden crayfish.

When we spotted a few boulders protruding from the water's surface, we stopped near the riverbank and I suited up for a hookah dive. Rather than breathing air stored in a tank on my back, I would receive my air via a long hose connected to an air compressor in the boat. While I dove, my colleagues would keep the boat near me and monitor the air compressor to ensure that I had a steady air supply. I geared up with a buoyancy compensator (BC) vest whose air volume could be adjusted to help me stay at a certain water depth, gloves to protect my hands from sharp rocks and shells, a handheld net and a mesh bag to collect and hold crayfish, a flashlight to see into dark crevices, a knife to cut away anything that might entangle me, and a camera to document the animals and underwater habitat.

We turned on the air compressor, which rumbled like a lawn mower as it filled the air hose. Minutes later, a whistle like a tea kettle signaled that the hose was full of air. I jumped into the water with a splash. As I released air from my BC and slowly sank below the water's surface, I was overcome with the same sense of serenity I always felt while diving.

Sinking until I hovered over the riverbed, I tried to avoid disturbing the sediments and further clouding the water. At best, I could see about ten feet in front of me through the dim, murky water. Eagerly, I moved toward a nearby boulder that looked as immovable as a building and pushed against it with my shoulder. Well, what do you know? Little ol' me was able to roll it over. Feeling like I'd been granted a superpower, I pushed over boulder after boulder, realizing that the current and the water conspired to make the rocks maneuverable. But with each boulder I rolled, the water became murkier.

Finding and catching the crayfish was a lot more difficult than I'd expected. Often covered in sediment and remaining motionless, they blended in with the riverbed. As I lifted boulders, flipped smaller rocks, and rolled logs, I finally started to see them. But when I reached out to grab my first crayfish, it jetted out of my reach with a quick tail flip.

Disappointed, I floated in place over the river bottom, searching for the crayfish's new hiding place. When I spied my target again, I slowly swam toward it and placed my net behind it, hoping that if I missed grabbing it with my hand, it would dart backward right into the net. But when I reached out to grab it, it shot away with another tail flip, evading both my hand and the net. With so many crevices for crayfish to use as cover, I often couldn't see where they had gone. But eventually, I began to get the hang of it and started to catch many of the crayfish I spotted.

Each time I caught one, I put it into a mesh bag attached to my waist. Glancing back to check on the bag, which floated near my legs as I swam, I spotted a large bass lurking behind me. Returning to my search, I turned over a few more boulders and missed a few more crayfish before finally catching another one. When I put my latest catch in the bag, I did a double take. More fish were behind me now—both bass and catfish. Four or five of them. All watching me.

The *Jaws* theme music suddenly played in my head.

Why were they watching me? These fish are known crayfish predators, so I quickly checked my mesh bag to see if the fish had somehow managed to steal the crayfish I'd collected. To my relief, all of my crayfish were still in the bag. I swam a little farther upstream, then glanced back again nervously. The fish were still there! My serene crayfish sampling felt decidedly less serene. *What did these predators want?* I was in their territory. *Maybe it was time to call it a day and head up to the boat.*

No. I reined in my uneasiness and focused on the task at hand. *These fish were not pursuing me . . . were they?* Shaking my head, I dismissed the thought and continued my search. But as I swam, I couldn't help looking over my shoulder. The gang of fish followed slowly in my wake. *They were stalking me!* I tried to reason with myself. Of course, the fish couldn't hurt me—or so I hoped. Staring resolutely forward, I continued looking for crayfish. As I collected another and reached back for my bag, I saw that the number of fish had doubled. Menacingly, they hovered about two arm lengths away, watching. I wondered if they were going to attack me. I had my knife and was prepared to defend myself if it came to that. Forcing myself to concentrate on the task at hand, I turned my back on the fish mob and pressed forward.

Another big boulder loomed in front of me. Flipping it over, I saw a large crayfish. My heart beat faster. If I could sneak behind it, I might be able to catch it on my first try. I moved into position, slowly put my net behind it, and reached out quickly to grab it or scare it into my net. But with a flip of its tail, the crayfish shot away. As it darted off, a bass to my left bolted forward and sucked the entire crayfish into its mouth tail first. After swallowing it in a single gulp, the fish immediately returned to my side. If I hadn't been looking directly at the crayfish, I would have missed an incredible interaction. It happened *that fast.* To my surprise, the large crayfish with its huge claws was no match for the bass, which showed no sign of having just consumed an unwieldy meal.

Now I understood why the fish seemed to be stalking me. These opportunists had realized I was uncovering their prey and were patiently waiting for my misses. *I was providing them with breakfast!* A column of air bubbles exploded upward as I burst out laughing. I had often read about crayfish and fish interactions, but I didn't know of anyone who had seen this type of encounter in a natural environment. And I couldn't help but feel a bit relieved that it hadn't been me the fish were after. That was the only crayfish I saw become an easy breakfast that morning, but I'd witnessed something remarkable.

As I swam to the boat, I could still see the bulging eyes of the fish that had quietly stalked me, using me to secure a meal. In the span of one dive, I'd gone from feeling calm and peaceful to nervous and paranoid and then amused and joyful. I left the water with a big smile on my face, wondering what awaited me on my next diving adventure.

Adrienne Richardson

Zanethia Barnett grew up in Estill, South Carolina, where she developed a love for nature. She spent many days fishing with her dad, hunting for shells on the beach, and crabbing in the creek behind her grandad's house. Her passion for nature pushed her to further her education, and she earned a bachelor's in agricultural economics from North Carolina Agricultural and Technical State University and then a master's in interdisciplinary ecology from the University of Florida, where her interest in aquatic organisms grew. During that time, she also became a scientific scuba diver, and diving became one of her favorite activities.

Zanethia's underwater explorations sparked her interest in studying crayfish and assessing how human disturbances affect aquatic systems. She focused her doctoral research on crayfish and became the first black woman to earn a PhD in biology from the University of Mississippi. Zanethia is a research fisheries biologist for the US Forest Service. Outside of research, she is passionate about exposing minorities to aquatic sciences and creating a more diverse, inclusive, and equitable field.

Moonlight Parade

Gina Piazza

Twenty-four hours may not seem like much time. But if you are tasked with staying awake for that long while perched in a tree in the middle of an African savanna, the hours just might be filled with transformative experiences. Inspired at a young age by my father, who fascinated me with stories of his many adventures as one of the first Peace Corps volunteers in Nigeria, I always knew I wanted to visit Africa someday. So when I discovered a study abroad program that gave college credit for volunteering with a cheetah conservation organization, I immediately signed up.

After two ten-hour flights and a long truck ride, I arrived at the organization's headquarters outside the city of Otjiwarongo in north-central Namibia. Here, overlooking the Waterberg Plateau, I would help with their day-to-day cheetah conservation work and conduct my own research on the impact of grazing on carnivore communities. I would also have opportunities to learn other field research techniques.

On one occasion during the full moon, a local student from the nearby town of Okakarara and I were tasked with observing a water hole for a twenty-four-hour period. From our perch up in a tree blind, we would count animals and observe their behavior. Our data would contribute to baseline information on local wildlife diversity. Late in the morning, David and I were dropped off in what seemed like the middle of nowhere. Reddish dirt and acacia thickets extended as far as the eye could see, except for a clearing with a large pond about 150 feet away. My professor, who had driven us to the spot, pointed to a nearby acacia tree. In it was a blind that looked like a roofless tree house, consisting of a wooden platform and sides with a long horizontal gap in the middle for a window. After announcing that the blind would be our home for the next twenty-four hours, my professor assured us that he would be back the following morning to pick us up.

Armed with books, notepads, snacks, water, and blankets, David and I ascended the rickety ladder to our blind and nervously watched our lifeline drive away. This was the first time I had been practically alone in the African bush with no way to communicate with the outside world. We were just two defenseless strangers in a tree, hoping everything would be okay for twenty-four hours.

Camping out in a tree blind in this semi-arid landscape took some getting used to. As the morning progressed, I spent my time swatting flies, trying to keep ants and corn crickets from climbing on me, and hoping no aggressive wasps lurked nearby. Whenever I left the tree for bathroom breaks, I checked to make sure scorpions hadn't sneaked into my shoes. Beyond coping with the relentless insects, we also needed to be mindful of venomous snakes—and, of course, large predators. But after several hours of tallying only warthogs scratching their behinds on trees, my initial anxiety faded.

As the hot afternoon hours crept by and the quiet enveloped us, David and I spent some time getting to know each other. Using a karroo thorn as a toothpick, David asked me what a childhood was like. I was startled by his question and found it surprisingly hard to explain what it was like to just be a kid. David had started working in the fields as a toddler, as soon as he was first able to stand. He was severely allergic to one of the crops his family was growing and suffered daily with his eyes swelling shut. He also asked what snow was like and if I knew any of the Americans he had met or seen on TV, such as Tupac or Michael Jordan.

The mosquitoes began to swarm as the sun went down, but I was distracted by an epic African sunset that blanketed the sky with every vibrant shade of red, orange, and purple. Crimson-breasted shrikes landed right above our heads, lingering nearby rather than startling and flying away. As darkness settled around us, the advantage of doing water hole observations during a full moon became apparent as a bright silver veil of moonlight highlighted animal movements amid the surrounding stillness.

Then began a dreamlike flow of animals so extraordinary I later wondered if my memory had played tricks on me. Tucked into our blind, we could see animals moving, but because they couldn't see us, they were relaxed and didn't run away. We were just part of the setting—and part of nature. More warthogs scratched themselves on trees and then gradually wandered off. A small herd of zebras soon arrived at the pond for a drink. Half an hour passed, and several giraffes strolled toward the water hole. Awkwardly spreading their longs legs so they could reach the water, they stretched their necks down and drank.

David and I settled into our opposite corners of the blind and relaxed into the safety of our hidden home as another wave of animals visited the water hole. A small group of elephants ambled into view, raising their trunks as they approached the water. To my surprise, instead of pausing to take a drink as I expected, they suddenly picked up their pace and crashed straight into the water at almost a run. Spraying water onto their backs, the elephants engaged in a graceful, playful dance in the moonlight. Watching them, my eyes stung unexpectedly as I was moved to tears by the beauty of the scene. With every toss of water, I felt like my worries and the baggage I'd brought from home were being washed away. Despite my tears, I couldn't stop smiling.

The parade of animals arriving at the water hole felt unnaturally fluid, as though I'd found myself in one of Gary Larson's *Far Side* cartoons where someone was opening a cage behind the bushes and releasing animals one group at a time. Half an hour later, a family of baboons showed up. At first, they went about their business, milling around and surveying their surroundings. Suddenly, though, one of them turned around, stood upright, and pointed its arm directly at the blind. More turned toward us, and soon they were all jumping up and down, screaming and pointing at us. Some leaped into trees to get a better look, still pointing and screaming. The illusion of safety and comfort I'd felt, tucked into our hidden tree blind, instantly vanished.

We were exposed and acutely vulnerable in the African bush. I froze, not moving a muscle as I silently begged the baboons to lose interest in us. Feeling sure they were going to attack us, I started to wonder if I would survive the night or become dinner for a passerby. I could easily imagine myself as one of the animals I had seen a predator take down. All of my senses were fully alert, my primal instincts readying me for fight or flight. But after what was probably only ten minutes, the baboons calmed down and then eventually left. Our invisibility cloak returned and my heart stopped racing. Hours went by without any more animals appearing.

The stillness settled around us until eventually we heard a quiet rustling in the distance. Then, into the moonlight stepped an enormous lion, his glorious mane framing a regal profile. The sight took my breath away. I prayed he wouldn't hear my heart pounding. No cage or fence protected me from this powerful animal. He drank briefly, then continued on his journey. Only minutes had passed, but the moment would stay with me forever.

No animals appeared for several hours after the lion, as though the other creatures had given him a buffer, leaving this majestic beast to travel his range in solitude.

Time dragged. David fell asleep, but I entered a state of clarity and awareness that I had never experienced before. I opened the book I had begun back home and found that reading Daniel Quinn's *Ishmael* by moonlight in the tree blind gave me a different perspective on the novel. When reading it at home, I had felt guilty and trapped, knowing I was doing things that seemed necessary but were detrimental to the earth, like driving my car to work or using electricity to keep the lights on. But now, reading in the blind, I felt no guilt. I was not using any resources and, in this rare moment, was having very little impact on the earth. The flow of wild animals, my vulnerability, and the timeless setting had given me a fresh perspective about my place in the world. I had no authority over the land or its animals. I was part of it all. The slow passage of time and the parade of animals brought home for me that humans are not separate from the wildlife that surrounds us—and we are less in control of our environment than we think. For hours, I had been at the mercy of the wild, and yet I had never felt more content.

The cool, fresh breeze in the quiet morning was heavenly. Announcing a new day whose possibilities had yet to unfold, the sunrise painted the sky in another Crayola palette of reds, oranges, and yellows. The sun warmed me, and I felt relaxed and comfortable. Animals still moved around calmly, going about their lives, seemingly oblivious to our presence. I had rarely experienced such peace and tranquility. It was nearly impossible for me to fathom that neighboring Angola was at war.

Before long, my zen was interrupted by the reality of the day beginning to bustle around me. Flies landed on me with persistence. Corn crickets hummed continuously. Jackals laughed in the distance. A two-day-old sting I'd received from a wasp encounter throbbed. My stomach growled. My skin was burned, scraped, and smelly. Sand permeated my pores and all of my belongings. My hair was tangled. And yet, I felt truly lucky.

Over a period of twenty-four hours, I had learned that I thrive in challenging situations. I felt a heightened awareness, my senses and perceptions sharpened by being just another animal amid the wildlife that surrounded me. Just as the elephants had washed away the day's dust in the moonlight, my night watch had cleansed me of the stress and pressure I often carried. I felt calm and peaceful and was filled with an overwhelming love for the world that encompassed my family, friends, and strangers.

My professor never returned to pick us up because his car broke down. Instead, One-eyed Jack came and collected us with the donkey carriage a few hours after the appointed pick-up time. As we rolled away from our tree blind, I looked back at the water hole one last time, picturing its moonlight parade. Then I turned to face the future, determined to hold on to all I'd felt there for as long as I could.

Courtesy of Gina Piazza

Gina Piazza's fascination with biology began early. Peering through a microscope at pond water as a child in the Pacific Northwest sparked a lifelong curiosity about the natural world. That early interest evolved into a master's degree in conservation biology from the University of New South Wales, with fieldwork including a study of kangaroo home ranges in Australia and research contributing to a conservation plan for a culturally significant island in New Zealand.

Gina now works as a senior biologist based in the US, assisting with habitat protection and restoration projects. Her passion for observing animals in their natural environments has taken her from tracking cheetahs in Namibia's savannas to counting thousands of albatross nests on Midway Atoll to spending time with silverback gorillas on the misty slopes of Rwanda's Virunga Mountains. Ever in search of her next adventure, she is still fueled by the same sense of wonder that started it all.

The Stump That Stumped Us All

Brianna Mann

Hunting for the New Mexico meadow jumping mouse is a bit like trying to complete a scavenger hunt while running a marathon. On some nights, our moonlit mouse tracking in the White Mountains of Arizona was peaceful—short jaunts through the woods to locate the one or two of our radio-collared mice that graciously chose to be stationary, interspersed with brief snoozes in the work truck. On other nights, our collared mice acted like hyperactive little tyrants, racing half a mile one way and then darting back in the opposite direction. Like obedient subjects, we raced after them, darting back and forth in the dark—in waders.

Radio tracking helps to illuminate the hidden lives of animals, providing insights that make it impossible not to love the creatures you're studying. Tracking our mice, listening to them munch seeds at the tops of grasses, seeing them scamper across the water or climb up into an alder tree, we not only obtained information that was essential to our understanding of the species and its conservation, but we also came to understand, on a personal level, how special each animal was.

The New Mexico meadow jumping mouse is special for a number of reasons. For one thing, it's endangered. These tawny-coated, white-bellied little mice are dependent on the vegetation that grows along perennial rivers and streams. Teeming with forbs and wildflowers, these riparian areas provide essential habitat that is threatened, primarily by cattle grazing. The agile jumping mice are so adapted to stream habitats that they swim often and even pad across the surface of the water, as confident in a flowing stream as they are at the fragile tops of six-foot-tall forbs. The longer we tracked these animals and witnessed their fairy-tale cuteness and unique habits, the clearer it became that protecting them and the riparian areas they inhabit is vitally important.

On one of the nights we spent tracking our beloved endangered jumping mouse, my crewmates and I each followed two mice. It was a mellow night, with our collared mice more or less spending time in their favorite areas, making our work relatively breezy. To track each mouse, we waded in the stream and, when necessary, navigated carefully along the stream bank to avoid damaging the riparian habitat. We followed the least-destructive footpath to where the transmitter signal was loudest, then narrowed in on the direction of the beeps until we knew the exact thicket in which our mouse was lingering.

By then, I had been tracking Mouse Number 5 for at least a week. I liked to think I knew him pretty well. He had a favorite patch of ferns and a cluster of coneflowers that he loved to snack on. Occasionally, he'd throw me for a loop and run upstream a hundred yards or so—forcing me to slither under a barbed wire fence and clamber over dense downed wood in pursuit—only to scamper back to his favorite ferns ten minutes later. And then, that night, in the span of less than an hour, he disappeared.

Jumping mice tend to stay close to water, almost always within ten yards, but we scoured the stream in both directions and received no signal. Our little mouse was nowhere nearby. Could he have been captured by an owl? How else could he possibly have traveled out of range so quickly? Jumping mice are not very big, after all. But then I heard the faintest signal—uphill. Away from the water. Much too far away.

The confusing thing about radio telemetry is that the signal emitted by an animal's transmitter can bounce off or be redirected by large objects in its path. For example, the radio signal can ricochet off trees, rocks, and steep slopes. If power lines are nearby, the static interference they sometimes create can make it a nightmare to decipher a signal's direction. But if the signal is there, the collar is somewhere nearby—maybe not in the exact direction you're facing, but somewhere strong enough that you can find it if you are determined enough. My crew was determined enough. Four of us combed the hill, chasing bounce-back from the pines again and again. Finally, we converged on a large, hollowed-out stump jutting about ten feet into the air.

Jumping mice are not big climbers. They stick to riparian areas and live in low, dense brush. This stump was not big enough to house an owl or hide a bobcat, but it was far too big to shelter a petite jumping mouse. There were no forbs or enticing jumping mouse snacks anywhere nearby. We were flummoxed. The chase had taken all night. Perhaps a raccoon or another predator had preyed upon our poor unsuspecting rodent. We returned to camp, utterly perplexed, saddened and expecting never to hear from Mouse Number 5 again.

The next night, Mouse Number 5 was under his favorite patch of ferns. He munched on his coneflowers. He slept in his day nest. And the stump? That mystery will haunt my dreams for years to come.

Courtesy of Brianna Mann

Brianna Mann grew up in Petaluma, California, and has spent her adult life working with wildlife in roles including managing natural history collections, surveying freshwater springs, and monitoring endangered species across the West. She works as a habitat restoration specialist for Bat Conservation International. Fieldwork has been a staple of her career, and the opportunity to see conservation happen on the ground continues to motivate and inspire her.

Brianna lives in Flagstaff, Arizona. In her spare time, she enjoys all things outdoors, reading, writing, and dancing, as well as a good local brew.

Just Keep Going

Shannon White

I never imagined I would pursue a career in field biology. But as my freshman year in college was winding down, I received a fateful email from my advisor asking if I wanted a paid summer research fellowship. No doubt most students would have thought that the project sounded idyllic: hiking and camping in Virginia's beautiful George Washington National Forest while studying native brook trout populations in Shenandoah Mountain streams. I had never been much of an outdoorswoman and didn't even know what a trout was. However, the project my advisor described sounded infinitely better than spending another summer working at the dog kennel, so I immediately accepted his offer.

Aside from having to adjust to camp life, I could not have had more fun during my first summer of fieldwork. From a science perspective, the research project was a total failure. We were attempting to study why trout move long distances, but the fish we were tracking moved less than fifty feet all summer. Nevertheless, I became enamored with mountain stream ecology. And long days of hiking were rewarded with campfire camaraderie and tasty meals prepared by my advisor, who had clearly refined his camp stove recipes through decades of practice. That summer left me wondering why everyone didn't pursue a career in field biology, and I didn't hesitate to reenlist the next summer.

It was during that second summer that I realized you have to be a little crazy to become a field biologist. Given what I had learned from our mistakes the first year, my fellow undergraduate researcher, Josh, and I chose a different research question. This time, we wanted to understand whether trout learned a new behavior—in this case, feeding on a novel prey item—more quickly when in the presence of other trout that had previously learned the behavior. If so, it would suggest that trout can adopt new skills and learn about their environment simply by watching the behavior of other trout. This process, known as social learning, is considered a fairly advanced cognitive strategy and was once thought to be exclusive to humans. But having spent a lot of time observing brook trout, my advisor was confident that these fish were capable of social learning.

Our general study design was simple. First, we would mark each fish with a unique combination of external ribbon tags—pieces of thin colored plastic threaded under the top layer of the fish's skin using a small needle. The tags were just large enough to allow us to identify individuals and observe their behavior if we sat on the stream bank with binoculars. Next, we would train fish in some pools to feed on canned mealworms. Surprisingly, trout that have never seen mealworms don't know that these worms are food. In fact, it takes trout about fourteen days (and a lot of taste tests!) to learn that they can eat mealworms. But once they do, they generally ignore other food items and consume only mealworms when they're available.

After fourteen days, we would move the trained fish to new pools in the stream and see how they influenced the feeding behavior of untrained fish. If, as we hypothesized, trout use social learning, we expected to see untrained fish feed on the mealworms after only a few days rather than the usual two weeks.

To document the time it took for fish to begin eating mealworms, and then how quickly untrained fish learned to consume worms in the presence of trained fish, we would monitor the behavior of tagged fish all day, every day, for about a month. The tagged fish were in multiple pools. We would observe each pool for thirty minutes four times a day, taking meticulous behavioral notes about every tagged fish we could see. If a fish looked at, ate, spit out, or fought to get a mealworm, we would note it. After each thirty-minute observation period, we would hike to a new pool and begin a new round of observations, continuing until the sun set over the ridgetop and darkness obscured our study subjects.

As with many field projects, what seemed feasible on paper posed serious logistical challenges in practice. Namely, this study required a lot of equipment, plenty of time to capture fish, and extensive data collection in an area that spanned about a mile of stream—and it needed to be completed in nine weeks. After a week of gathering materials, building fish feeders from scavenged PVC tubing and electrical components, and finalizing arrangements, we headed back to the forest where I'd worked the previous summer and set up our tents. We knew the first two days of the summer field season were going to be the most physically demanding.

We devoted our first day to installing equipment and made continuous high-speed circuits, hauling sheets of plywood and plastic piping up the mountain to an installation crew and then scurrying back to camp to retrieve more supplies. By the end of the day, our crew had installed twelve enclosed observation platforms. These platforms rested—sometimes quite precariously—on the side of the cliff bordering the stream and allowed us to observe trout behavior (using binoculars with polarized film over the lenses) without the fish seeing us (important since trout have great eyesight and spook easily). At each pool we'd also installed a fish feeder by bolting it onto rebar driven into the stream bottom. We finished installing the last platform by the light of our headlamps. The campfire was eerily quiet that night as we passed around the Tylenol and contemplated the inevitable onset of aching muscles.

As the sun rose the next day, we could hear each other's sleeping bags rustle, but no one seemed eager to leave their tent. This was unorthodox behavior for this crew, a group of early risers who generally had coffee brewing by 6 a.m. I think most of us were unconvinced that our legs were capable of functioning. But one by one, each of us eventually crawled out of our tent and prepared our mind and body for the day ahead. Soon after, we put on our waders, devised a game plan, and headed down the trail toward the downstream end of our study site. We knew we were facing another battle against time because we needed to catch and tag all the fish for our study before making the three-hour drive back to campus.

Thankfully, capturing the fish provided a bit of a reprieve. The cool water helped silence our screaming muscles, and the excitement of electrofishing—using an electric current to briefly stun fish so we could capture them—made the time pass quickly. We released the last fish back into the stream as the sun dipped below the tree line and felt relieved that we had accomplished everything on our ambitious to-do list.

Eager to get home, we decided to take a shortcut to camp that required hopping between big boulders that bordered the stream. In the back of my mind, I knew this was a terrible idea since I was a good foot shorter than any of the other crew members. But being vertically challenged had forced me to get creative while doing fieldwork before. I readily agreed to the plan, and we donned our thirty-pound packs and set off. Within seconds the others had built a decent lead, while I lagged behind, trying to find my own path through the rugged terrain.

That's when it happened. I suddenly heard a loud pop and collapsed to the ground in blinding pain.

My advisor ran back to me, asking what had happened. I mumbled incoherently for a few seconds before managing to convey that I had hurt my knee. My leg was twisted under my body, and after several minutes I mustered the courage to ask my advisor to straighten it so we could assess the damage. Surprisingly, I felt no pain. A short while later I stood up, put a bit of weight on my leg, and again felt very little pain. *Phew, maybe it was just a sprain.*

Daylight was fading quickly, and no one had their headlamp with them. So we agreed that the rest of the crew would head back to camp as fast as possible to pack everything up, and I would slowly make my way to the car. As their shadows disappeared into the woods, I took a few small steps and eventually worked up to full strides. After a few minutes, my limp was hardly noticeable, and I started to regain my confidence. But suddenly my knee buckled, and I found myself lying on the forest floor again, writhing in pain.

Despite my misery, my overwhelming concern soon shifted to our ability to complete the project. I knew that someone from the crew would be back if I needed help. But the entire field season depended on my being able to hike through the woods, and I knew I needed to start by making it back to camp on my own. Wedging my body between two rocks, I used my arms to pull myself up to a seated position. After a few minutes, I stood and thought to myself, *Just keep going*, knowing that I could figure out the rest later. Those words ended up becoming the mantra I have repeated throughout my fieldwork career to grapple with everything from broken-down vehicles to getting lost in a rainforest. But on this day, repeating these words got me back to the car just as the crew finished loading our gear.

Over the weekend, my knee was a little swollen, but I felt like I had dodged a bullet and had no major injuries. I was the first to arrive at the Monday morning planning meeting with my advisor and Josh. After they let out a collective sigh of relief that I was walking, my advisor dropped the bomb that, unlike the previous summer, he wouldn't be able to help us with most of the fieldwork. The silence that followed his announcement was deafening. The fate of the project now rested entirely on the shoulders of two undergraduates who felt wholly inadequate and unprepared for making major decisions about the research.

Still reeling from the shock of this announcement, Josh and I met at six the next morning to start the first two-week phase of the project. We hadn't bothered to check the weather forecast, since we assumed we'd experience all possible weather scenarios. As it turned out, we experienced little more than rain for our entire month of camping. Fortunately, we could do our behavioral observations in all but the heaviest downpour. But finding anything dry at camp soon became impossible. Starting a fire was challenging, too, so our meals consisted mainly of granola bars and whatever we could heat quickly on a small camp stove. Electronics, no matter how carefully stored, began failing after just a few days. Our clothing was permanently damp, and mold grew on every surface. Because air temperatures topped 90 degrees every day, we abandoned our waders and just walked through the stream in boots and long pants. But after spending hours each day hiking in knee-deep water, we started developing something akin to trench foot.

Keeping the study running also presented unique challenges. The feeders became hot spots for minks and raccoons, which would steal a week's supply of worms in one night. In addition, we frequently found the feeders thrown several feet into the forest and the rebar that anchored them bent in half. We deduced that this chaos was probably the work of black bears, which only became disconcerting when we realized that the pool with the most heavily damaged rebar was about forty feet from our campsite.

The near-daily equipment tampering added to the tedium of data collection. Josh and I typically left camp around 7 a.m. to start our behavioral observations. We divided the study reach and each observed the pools in our respective areas four times daily, so we spent our twelve-hour observation days alone. On the best of days, I appreciated the opportunity I'd been given to witness the complex social dynamics of an iconic fish species. On other days, I felt disheartened by the monotony of the work, the long hours of isolation, and the less-than-ideal environmental conditions.

At the end of day fourteen, Josh and I sped away from the field site knowing that we had successfully trained twenty brook trout to feed on mealworms. After a weekend off, we returned along with our advisor to move the trained fish to new pools. Everything went smoothly. By early evening we had moved a few trained fish to each of six pools, and my advisor was packing up to head home. Before he left, he gave us a long pep talk about how our observations in the next two weeks ultimately would tell us whether the data supported our hypothesis that trout engaged in social learning. With any luck, in just a few days we would see that the untrained fish had learned from the trained fish and had started to eat mealworms.

After our advisor left, Josh and I parted ways to do our initial observations. During the first round, we confirmed that most trained fish were still in the pools we had moved them to a few hours earlier. Before leaving each pool, we opened up the feeders in preparation for our next set of observations. As I hiked back to start my second round of observations, I worried that the trained fish might be so disturbed by their move to new surroundings that they wouldn't eat the mealworms at all. When I got to my first pool and peered into the stream, my heart sank. But for a different reason. Every fish—trained and untrained—was feeding on the worms. Since we had just restarted the feeders a few hours earlier, it made no sense that the untrained fish would be feeding this quickly.

Maybe it's a fluke, I thought, perplexed. I waited for the feeder to drop another worm and looked again. All of the fish attacked the worm. This couldn't be right. Something seemed to have gone awry with our experiment. After thirty minutes of watching worm after worm get devoured by untrained fish, I was relieved to move to the next pool, where I hoped to see the fish behaving as expected. But within seconds I realized that the same thing was happening here. Not only were untrained fish feeding on the mealworms, but they were also actively fighting off the trained fish to access the food.

Watching in bewilderment, I felt sure that the project was doomed. Something inexplicable had happened, and we were wasting another summer on another failed project.

Back at camp, Josh revealed that he had seen the same thing. We speculated on what might have happened, but no hypothesis we formulated could explain why the untrained fish were already feeding on the mealworms. We had no cell phone service to call our advisor and we knew it was too late to change the project, so our only options were to continue the study as planned or quit. We decided to carry on, carefully observing the fish and collecting the best data possible for thirteen more days even though it felt pointless. So through the rain, daily equipment failures, and my progressively worsening knee, we just kept going. At the end of those two weeks, we drove away feeling completely defeated by the scientific process.

Back on campus, our advisor encouraged us to enter our data quickly so he could statistically analyze them while he was away on a trip. Sitting at the computer, still exhausted from a month of camping and fieldwork, Josh and I couldn't help but commiserate about the futility of entering pages upon pages of worthless behavioral data into a spreadsheet.

A few days later, our advisor convened a meeting to discuss the results of his analyses. Before we even walked into his office, he shouted down the hall, "You're not going to believe this. It worked!"

For the next couple of hours, he shared the results of his statistical analyses, which showed us what our eyes had tried to tell us all along: the untrained fish *had* learned from the trained fish, mimicking their feeding behavior almost instantaneously. The process had happened so much more quickly than we had expected, and Josh and I had been so focused on the project details that had not gone as planned that we hadn't even considered the possibility that our hypothesis might have been supported by our data.

Today, more than a decade later, I still wonder how it took a series of graphs and statistical analyses to convince us that what we had observed was true. Perhaps, as undergraduates overwhelmed by our first independent field project and the unrelentingly miserable conditions, we didn't yet appreciate that the scientific process often proceeds in a nonlinear fashion. The path from research question to data collection to data analysis to final results often takes unexpected twists and turns—and sometimes you just have to trust the process. After conducting many more field studies, I've learned that I almost never feel confident while collecting data that they will prove to be useful. But I know now that quality data usually tell a story, even if it's not the story you initially set out to tell. And even though it can be difficult to persevere when nothing seems to be going as expected, in the end you'll rarely regret the hard work you've invested.

Oh, and my knee? Turns out it was more than a sprain. I had torn a few ligaments and a meniscus, though I didn't find that out until almost nine months later, at which point I delayed surgery because I was planning a field season in Colorado.

And the study? It's now published, and it's still one of my proudest research accomplishments.

Catherine Noyes

Shannon White grew up just outside Richmond, Virginia, without any plans of pursuing a career in environmental science. That changed with a single undergraduate class at Randolph-Macon College that sparked a lifelong passion for aquatic ecosystems. She went on to earn a master's degree in fisheries and wildlife science from Virginia Tech and a PhD in ecology from the Pennsylvania State University.

A research biologist at the US Geological Survey's Eastern Ecological Science Center, Shannon studies aquatic species and ecosystems across a broad range of habitats—from small Appalachian streams to the open Atlantic Ocean. Her research spans conservation genetics, acoustic telemetry, and population ecology, all aimed at informing the management and recovery of imperiled species.

The Rock

Deanna (Dede) H. Olson

"Dede means 'dead' in Sranan Tongo," he said as he handed me my letter. A twinkle in his eye and my wide-eyed gape at his reference to my nickname made us both burst out laughing. It was Christmas 1980, and the first mail had arrived at our forest camp at the Raleigh Falls / Voltzberg Nature Reserve in Suriname, a small country on the northern coast of South America. Sranan Tongo—a Creole blend of Dutch, English, and other languages—is the day-to-day language spoken in Suriname. I was in the country with my boyfriend, Mike, to help two graduate students, Kurt and Pepper, with their research. The mail from home was the best Christmas present ever for the four of us at our isolated camp—a simple thatched roof held up by tree poles, where we'd just spent the first few weeks of our six-month sojourn.

My letter had made the same journey as Mike and I had, by air from San Diego to Miami, then to the island of Trinidad before heading to nearby Guyana, and finally on to Paramaribo, the capital of Suriname. Our mail went to the Paramaribo headquarters of STINASU, Suriname's Foundation for Nature Conservation, which coordinated the researchers at the reserve. STINASU had entrusted my letter to a group of tourists who were embarking on an exhilarating all-day trip: first by van, spewing a plume of red dust along the dirt road from the city to the Coppename River; then by motorized dugout canoe upriver to Fungu (pronounced *foongoo*) Island. This ride is etched in my memory. Powering upriver in a hollowed-out tree into a pristine tropical rainforest felt like being in a story from National Geographic, which had, in fact, funded our expedition.

Upon first entering Suriname's interior, the mailbag-toting tourists had passed walls of greenery three hundred to four hundred feet high along the river and may have glimpsed some of the area's 470 bird and 8 primate species. Careful observers had also seen caimans in the water and, if lucky, a giant river otter. I remember the ride as an adrenaline rush of the senses, with new and exotic smells, sounds, and sights: the floral to earthy aromas of the humid air that enveloped me like a heavy, warm, moist blanket; the clusters of raucous parrots congregating at mud banks to pick up minerals essential to their diets; and the cacophony of oropendolas—big conspicuous birds in the blackbird family—attending their long, pendulous nests that dangled from trees over the water. The river was the entrance to an Eden sparsely inhabited by people.

While overnighting at the spartan barracks on the mid-river Fungu Island, the tourists had enjoyed the evening spectacle of cane toads feasting on clouds of moths attracted to a generator-powered night light. Fungu Island was home to the forest rangers and their families, who, encircled by a natural river moat, seemed sheltered from potential jungle dangers. The rangers were the descendants of formerly enslaved Africans who had escaped and taken refuge in the forest interior back in the colonial era when the Republic of Suriname was known as Dutch Guiana. Savvy about remote forest and river life, the rangers I saw were expert boatmen, and some had in-depth knowledge of the local flora and fauna.

The next morning, the rangers had transported our mailbag as they led the tourists on the day's outing. They had first canoed from Fungu Island across to the Coppename River's shore, and from there, they had hiked to our forest camp at the Voltzberg granite dome, the signature monolith of the nature reserve. And now this amiable and teasing ranger was handing a letter to the "dead" girl in the jungle! I was twenty-two years old that Christmas at the Voltzberg. Mike and I had recently graduated with our bachelor's degrees in biology and had been hired as field technicians for a study of the breeding behavior of the Guianan cock-of-the-rock.

Large neon-orange birds, male Guianan cocks-of-the-rock have semicircular mohawk-like crests edged in black extending from their bills back to their napes. Streamer-like feathering in their wings adds a flourish when the males show off—posing with a certain self-aware hauteur—to the drab brown females during courtship rituals on communal display grounds known as leks. As with many other bird species, the females choose their mates, leading to fierce competition among males to woo females. In lek-breeding birds, males assemble on the display grounds and put on simultaneous feathery shows for visiting females. Depending upon the species, displays can include demonstrations of prowess, where males sing and dance, or beauty contests, where males—like our candy-orange dandies—pose as if modeling the latest must-have attire in storefront windows.

Upon arrival of a female at the lek, male Guianan cocks-of-the-rock give a *hey* call and drop from the tree canopy to the ground, their fluorescent plumage contrasting with the backdrop of the forest floor. There, each male strikes a pose on a three-foot-diameter patch of ground that he has cleared of vegetation and debris to create his own personal stage. In my mind's eye, I can still picture fifty males *hey*ing and dropping in a downward flow of feathers to their courts, arranged three to four feet apart in a lek about the size of a basketball court. Once on the ground, each male remains still but cocks his head sideways, flaunting his crest while keeping an eye on a visiting female. Flitting over the crowd of posing popinjays, the females move through the trees overhead, comparison shopping. After spotting the male that most appeals to her, a female drops down to his court, and the two mate.

Another of the region's lek-mating birds, the white-bearded manakin, caught our attention as well. Because Mike and I were essentially a twofer, applying as a team for one field technician position and thereby contributing twice the work, we were given time to do our own research. Studying the lek-breeding birds was irresistible. Manakin males are small fistfuls of fluff, stark white with a jaunty black cap and cape. The females are drab, camouflaged in olive green so as not to draw attention to themselves when on their nest. Like their lek-mating orange counterparts, the male manakins drop from the trees whenever a female approaches and display on a small court, one to two feet wide, that each bird has cleared on the forest floor. But the similarities between the two species' mating behaviors end there.

White-bearded manakin males cluster their display courts in areas of dense saplings. Each male strips saplings of leaves and twigs to delineate his court's boundaries. The males then attract females not by posing but by conducting heroic acrobatics, jumping like popcorn from a sapling stem to the dirt floor and back to a stem, while producing a rapid percussion of *click*s, *snap*s, *whirr*s, and *chirrup*s—a gymnastic pole-and-floor routine. With their white plumage contrasting against the green forest backdrop, the males are like whirligigs in motion, providing a fireworks show with a castanet-laden symphony.

At the main manakin lek that we studied, up to five males displayed at once: a five-ring circus. The action could last a minute or two, or it could be shorter, punctuated by a single female dropping down to a court, and in a blink, a mating, then a flutter as the female appeared to compose herself and then departed. The manakin and cock-of-the-rock studies both aimed to decipher the hows and whys of the birds' mating systems in this unusual environment.

The unique geology of this area features the Voltzberg granite dome—The Rock—and associated smaller domes and rock outcroppings, as well as flat, exposed rock plates scattered through the forest. The Rock—a nub of solid rock eight hundred feet tall that looks like the tip of a giant's thumb jutting out of the ground and extending about five hundred feet above the forest canopy—emerges out of a sea of treetops that extends to the horizon in all directions. A tourist destination and national treasure, it now graces the back of Suriname's twenty-dollar bill, is part of the Central Suriname Nature Reserve, and was designated as a UNESCO World Heritage Site in 2000.

The rock outcrops of the Voltzberg area are also ecological islands, harboring unique floral and faunal communities. The area's exposed domes and plates radiate equatorial heat and repel penetration by plant roots. Desert-like cacti find purchase in occasional cracks amid rocky expanses, and unusual, assorted scrubby plants merge into the jungle along the edges of the granite extrusions. A pineapple, growing from the rock plate near our camp, was one of the unexpected fruits we encountered. We had a monotonous diet with no access to fresh fruits or vegetables, so we eagerly awaited the fruit's ripening, but alas, our harvest was hijacked by a large tortoise. With a *hiss* it gave us fair warning: steer clear and finders keepers. The tortoise was loaded with ticks, which made our deference easy, and we figured it needed the sustenance more than we did.

Along the boundaries of the Voltzberg area's rock plates, emergent patches of forest composed of young saplings or older, stunted trees provide perfect habitat for white-bearded manakin leks. Rock outcrops are essential for cock-of-the-rock breeding, too, as befit the bird's name. Female cocks-of-the-rock build mud nests near the domes in protected caves and along rock walls. Males establish their main leks near these nesting sites. The area provides an ideal mix of conditions for these two intriguing species and a host of others.

While focusing on our study animals, we immersed ourselves in this unique ecosystem. As part of Kurt's research investigating the birds' diets, including what cocks-of-the-rock feed on during the breeding season, we collected and sorted the droppings they left at their display courts or under their nests. My advice to anyone working with bird droppings in the tropics is to *not* procrastinate and let the excrement sit in a baggie for any amount of time before dealing with it. We discovered the hard way that the droppings readily decompose to a slurry that roils with unknown creepy-crawlies and exudes a grotesque, rotting stench!

To further help Kurt in his research on the birds' foraging ecology and behavior before, during, and after the breeding season, Mike and I also tracked radio-tagged individuals. Since radio-tracking technology was in its infancy at that time and ready-made radio transmitters were not available, our transmitters were built to order by Kurt. His ingenious design used the smallest watch batteries available, dental epoxy for rainproofing, and hand-sewn cotton backpacks to temporarily affix the transmitters to the birds. Cotton rots quickly in those jungle conditions, so the transmitters would fall off once they were no longer useful.

We caught the birds in mist nets—fine-mesh nets that we strung through the forest near their leks or nests—and carefully outfitted them with the radio-transmitter backpacks. We then tracked the birds to determine where they were feeding. Because we were more likely to detect the radio signals from high points, we established three elevated radio-tracking posts around the main lek and nesting areas. Two of the posts were atop isolated rock outcrops that protruded into the forest canopy, and the third was atop the Voltzberg dome. We named the farthest midcanopy rock outcrop Olsonbergi because that was *my* main designated post. These high outcrops were draped with tree roots and lianas that we used as climbing ropes to reach the top. Once atop a rock in the midst of the tree canopy, I felt like I was part of daily jungle life.

From these perches, I experienced wondrous moments and comical contradictions. Pairs of beautifully colored macaws flew gracefully overhead, sometimes visible through gaps in the canopy, while emitting almost-painful screeches like those I imagined a pterodactyl might make. Strikingly hued toucans, looking oddly off balance with their enormous bills, animated the forest with an energetic *schwhoop schwhoop schwhoop* of their wings as they struggled to gain altitude, followed by the *sssshhhhwuu* of steep, gravity-assisted descent as they navigated through the branches in wavelike undulations. Despite their bright plumage, I always heard each of these species before seeing them.

Being alert to sounds opened a door to another world in the midcanopy. At Olsonbergi, a *whirr* once drew my attention to what appeared to be a six-foot-wide swarm of insects—maybe flies or beetles—swooping down and alighting on a nearby branch. Within seconds, the insect cloud shrank into itself, condensing to a single solid mass—a two-foot-long, rugby-ball-shaped hive whose bimodal color pattern suggested that each bug was black and white. The next day the hive was gone.

On another day, as I listened for radio transmitter signals from my high rock, I saw a troop of Guianan red howler monkeys coming straight toward me. They were known for continuous group howling—a roaring chorus that could be heard miles away, especially near dawn and dusk, but this troop was on a midday move. They came along harum-scarum, creating a racket and interacting with other monkeys as they traveled. When they were about twenty-five feet away, several individuals suddenly spotted me sitting quietly on my rock and stopped stock-still, then looked alarmed, squawked agitatedly, cocked their heads curiously, and finally ran back and forth interacting with each other. One large howler—perhaps a troop leader—was a focal point of many of the interactions. I quickly realized I should keep that bigger monkey in my peripheral vision while also keeping track of the bold youngsters. My startled mind raced. Would they attack?

Don't look directly at them, I thought. *That might mean aggression.* I wondered how fast I could flee by shinnying down the roots and lianas. I tried to stay quiet and still. Although it seemed like forever, it likely wasn't long before the troop started to move away. Soon they stopped eyeing me and resumed their traveling-troop hijinks. Only then did their unimaginable odor knock me sideways. Ah, what a wondrous yet stinky place this could be!

The top of the Voltzberg dome afforded the best radio-tracking potential since its height reduced signal interference from the canopy and other rock outcrops. If the birds flew far from the lek or if their signals were blocked at the lower posts, we could often locate them from atop The Rock. The dome had near-vertical sides except for one sloped area that provided our route to the top. The rough granite afforded traction, which helped me master my height fright while ascending that exposed track in my Vietnam-era army boots.

An unforgettable event on my first day up on The Rock made the perilous journey worthwhile. Sitting atop the Voltzberg, Mike and I heard a sound like a train whistle above us. Looking up, we spotted two unidentifiable dots careening out of the sky. Diving side by side, a pair of king vultures shot past us and barreled into the forest canopy below, a distant rustling of leaves the only indication of their passage. We wondered what had precipitated their stoop, which caused us endless speculation. We never saw king vultures again.

Later, another wildlife sighting from the top of the Voltzberg left me similarly slack jawed. After hearing a small sound below me, I spotted a brown-throated three-toed sloth ascending the vertical face of the dome. The species is adapted to moving through the tree canopy, but this free-climbing ascent looked like a hard-fought, bare-knuckled, soldier's belly crawl. Upon reaching the top, the sloth made its way over to a lone bush growing out of a crack and took up residence in its low branches. I was eye to eye with it! The sloth's black-and-white face markings gave it a permanent smirk. After relishing our face-to-face encounter, I hightailed it down the slope and back to camp in record time to see if anyone there had a camera. We scuttled back up the dome, camera in hand, and hooray, my sloth friend was still there. We soon learned that sloths are not always slow, as a lightning flash of razor-sharp claws swiped out at us when we got too close for its comfort. As I availed myself of the remarkable opportunity to see a sloth up close, I spotted a slug-like thing the size of my pinky finger living in its green, moss-tinged fur. *Ew! Maybe basking on the rock-top sauna was ectoparasite therapy?* The sloth was gone the next day.

In addition to tracking the birds to study their foraging ecology for Kurt, Mike and I also occasionally helped the other graduate student, Pepper, whose research focused on understanding cock-of-the-rock social interactions at the lek. Pepper had banded most of the birds with colored leg bands so he could determine the outcomes of interactions between individual males and females. But we kept seeing an unbanded lone male near some nesting caves. We needed to band him to see if he was also visiting the lek, where Pepper observed activities from a blind.

Pepper assigned me the task of mist-netting the rogue male, but since I'd never banded on my own, gave me a whistle to call for help if I caught the bird. Soon after my vigil began, I saw the lone male flying toward the net. With a *bam* he hit it and dropped upside down into a pocket of loose netting that had formed upon impact. Jumping to my feet, I whistled like a maniac. Out of breath, I paused, listening for Pepper's arrival, then waited. And waited. I whistled again. Then waited. Still no sign of Pepper. I didn't know the route to the lek from this location, so I couldn't easily jog over and get Pepper. The feisty bird was struggling in the net and would get hopelessly tangled or injured if I left him much longer.

Heart racing, I opened the supply box Pepper had left with me and perused the contents: leather gloves, metal band for one leg, pliers to close that band, color bands for the other leg, glue to secure the color bands. Pushing the items aside, I searched for the list of color codes that would tell me what color bands to use for the next bird to be banded. *It wasn't there*! Uh-oh.

I whistled again. Then listened to the deafening silence that followed.

Okay, take a big breath. I can do this, I decided. *Rather than putting the usual two color bands on one leg, I'll put three. That will be unique. There's enough room on the leg that it won't bother the bird, and I won't be using a color-band combination that Pepper has already used.*

Later, when Pepper returned to get me and take down the net, he seemed mystified that he hadn't heard my whistle—but he was glad the male was banded. And I learned I could rise to the occasion when things didn't go as planned in the field.

Working alone, far afield from our coworkers, was standard procedure for us. But we recognized that the Voltzberg had many potential hazards, including diseases and venomous animals like snakes, spiders, and insects such as wasps and army ants. Every day we'd share our plan and head off separately, being sure not to take undue risks while out, and reconvene at camp by the end of the day. I was unfazed by this arrangement, as I had been a free-range kid, often making day-long solo forays and returning home for dinner. Our final safety net was a nightly check-in via walkie-talkie at an appointed hour with the rangers at Fungu Island. The walkie-talkie worked like a charm—except when it didn't. Sometimes we simply failed to connect or connected with the wrong person. We never had an emergency at the Voltzberg that year, but I had one close call.

It was the day I went with Kurt to the Olsonbergi rock outpost for the first time in order to confirm its utility for radio tracking. He had found the location the previous year and had cut a trail through the forest understory. As we walked together toward the outpost, we kept our machetes out to clear away any annual growth—mostly fern fronds—that intruded on the trail. Once atop Olsonbergi, Kurt set up the radio receiver and dialed in the frequency of the test radio transmitter he had just made. The receiver's beeps indicated that the transmitter was working like a charm. He then stayed at Olsonbergi to track me as I took the transmitter along the trail back to camp.

Easy-peasy, and off I went . . . until the trail disappeared! I backtracked. The trail was quite evident. I moved forward. The trail disappeared. After repeating this several times, I finally just sat down on a log at the end of the known trail and waited for Kurt to find me as he returned to camp. I had the transmitter that he was experimentally tracking, after all, so I knew he would find me eventually. Hours later, as it started to get dark, my mind fixated on all the perils I might face out in the forest alone, especially the sand flies that came out at night and sometimes carried the parasite that causes leishmaniasis, a disease that produces skin ulcerations.

Back at camp, the three guys had convened and started dinner. "Where's Dede?" one of them asked. In the silence that followed, the three exchanged looks. Apparently, I had been tracked, as anticipated, until the receiver's battery died, and then Kurt had headed back to camp by a different route.

"I bet she's at that treefall that we had to cut the trail around last year," Pepper said as he grabbed a lantern and the three of them rushed off to find me in the gathering darkness.

Lions and tigers and flies, oh my! flitted through my mind as night rapidly descended. Finally, I heard, "Dede!"

"I'm here," I shouted back, relief washing over me. My three coworkers seemed to magically appear next to me. Flabbergasted, and a bit embarrassed, I stammered, "I lost the trail."

"It's here," Pepper revealed with a simple arm wave. We all looked, and from my side of the log, the trail was overgrown and invisible, appearing to be a dead end. Quick work with our machetes reopened the path, and we trudged back to camp. After a hasty dinner, we got into our hammocks, strung in parallel in our open-walled thatched hut, and settled in as their pendulum-like swinging slowly subsided. As always at night, we said "sreebi bun!" to each other (though I always pronounced it *sleepy boon*)—"sleep well" in Sranan Tongo.

My Suriname story unfolded with spectacular lekking birds as the raison d'être, but my personal journey during those six months was broader and laid a cornerstone for the next forty years of my field ecology career. Looking back on the day I got lost, I realize that while in the Voltzberg, I developed what has now become my standard approach to handling crises: control the panic and proceed logically, doing the best thing for that moment. A couple of years later, one of my fellow graduate students remarked on how well I weathered stress. "You're a rock," she told me.

The statement surprised me at the time. But recalling it now reminds me how formative my time at The Rock was. Working there instilled several important life lessons: delight in nature's ongoing operetta, keep occupied if alone, strive to overcome your fears, keep vital paths clear of obstacles, try to make the best possible decisions when decisions must be made, and cope with the unexpected with good humor. And at day's end, put everything aside and sreebi bun.

Little Running Paws

Sandra Lai

After circling twice at low altitude to assess the state of the ice, the small Kenn Borek aircraft finally landed on the frozen lake. Nearby, two large Weatherhaven shelter tents made up the base camp on the south plain of Bylot Island in the Canadian High Arctic—my home for the next few months. I was glad to finally arrive after almost three hours of sitting on a jump seat in a plane crammed full of carefully secured research gear, camping supplies, and food boxes.

As I emerged from the back of the plane, the sunlight reflecting off the powdery white landscape momentarily blinded me. It was late May. Although the sky was already brightly lit by a sun that would never set, snowmelt had not yet started. Everywhere I looked, deep snow blanketed the glacial valley where we had landed, and long streaks of snow enfolded the surrounding black mountains.

I had come to Bylot—a large uninhabited island located just off the northeastern tip of Baffin Island in Nunavut, Canada—to study arctic foxes. For decades, biologists on Bylot have examined how predators and prey interact within the arctic food web, providing key insights to guide conservation and management efforts. As part of that research project, I planned to fit the foxes with miniaturized satellite collars that would allow me to follow their movements remotely for at least a year—the life expectancy of the collars' batteries.

Being able to visually recognize each animal in a study area is crucial to many long-term wildlife monitoring programs, since resighting individuals year after year allows researchers to assess the animals' survival and reproductive rates. Therefore, we also routinely marked these foxes with small colored plastic ear tags to allow individuals to be visually identified from a distance. The ear tags had unique identification numbers, and their color combinations became the animal's "name," although we also bestowed nicknames on our favorite foxes.

As a first step, we needed to find and capture foxes in traps specially designed not to hurt the animals—leghold traps that had rubber-padded jaws or Tomahawk cage traps. Every year, the logistics of the operation were daunting. Our fox team consisted of four biologists tasked with surveying the 110 fox dens we had identified in our 230-square-mile study area. We visited each den twice during the summer, though only 10 to 50 dens were used by arctic fox pairs during any given year.

To cover the large distances between fox territories, teams of two people were transported by helicopter and deposited—along with their necessary camping and trapping gear—a few miles from an active fox den. Each team stayed at a location for five to six days, trapping day and night to capture the foxes from that den, and then was flown to another den.

When I first set foot on Bylot, I was unfamiliar with both extreme camping and fox trapping. My only other experience with remote camping was the three months I had spent in Tanzania with a UK-based conservation organization whose trips were geared toward adventure-seeking British youngsters on their gap year. But working in the High Arctic was my life's dream, and I was ready—or was I?

As it turned out, I was in for many surprises that summer. In June, I met my very first arctic fox—a young female still in her snow-white winter coat. We trapped her right before a heavy snowstorm hit the area. When I spotted her in our trap, I radioed my teammate, Cassandra, telling her where I was and asking her to bring the tagging supplies and a satellite collar. Then I hid a few dozen yards away from the fox to avoid causing her any additional stress. Fog was rolling in gently and thickening; I didn't know that a storm was brewing. By the time Cassandra arrived an hour later (after losing her way in the fog), I was getting chilled, but I didn't pay attention to the cold because I was focused on learning how to handle a fox correctly.

Often described as slightly larger than an average house cat, arctic foxes are small—but I didn't expect them to be *that* small. Of course, I had seen pictures of arctic foxes and read that they weighed between six and ten pounds, but reading about something and seeing it in real life are sometimes very different things. I frowned and hesitantly asked Cassandra, "Is this a cub?" Cassandra lifted her head, looked me straight in the eyes, and said, "No. This is an adult. When you encounter a cub, you'll know it's a cub." I was dumbfounded, but later that summer I would realize how right she was.

A few minutes later, the dense milky-white fog had surrounded us, and snow started falling lightly. I was now freezing, and we were rushing to finish recording data on our field sheet before the weather got worse. We were afraid the fox would get cold, too, so I secured her on my lap to get a better grip on her and to shelter her from the icy ground. Soon we were finished, and as Cassandra started packing up our equipment, I whispered: "I think the fox is okay . . . actually, she's the one warming me up!" In a few minutes, her plush coat had warmed me in a way that my three layers of clothing hadn't, giving me a firsthand experience of the warmest fur in the animal kingdom. We released her soon after, and she quickly vanished into the fog. We named her Tempête, the French word for "storm."

Tempête did not have cubs the year we first captured her, but I met her again over the next two years. The following summer, she had grown from a shy young female into a mighty mother of thirteen cubs. She was so fierce that when she was retrapped and put inside a capture bag, instead of just curling up like most foxes do, she stubbornly stayed upright, standing on stiff legs. In my mind, her fiery spirit reflected the incredible dedication with which arctic foxes raise their young during the short arctic growing season.

For five summers, I camped in small tents in the middle of Bylot's tundra, sharing the beautiful arctic scenery with the foxes. I watched them tirelessly hunting lemmings (small rodents that constitute their main prey), quarreling with snow geese and attempting to steal their eggs, and sniffing out the well-hidden nests of shorebirds and Lapland longspurs to feast on their little eggs. Except for migrating species that are heavily hunted after leaving the Arctic to reach their wintering areas, arctic animals show almost no fear of humans. Arctic foxes curled up to nap a few yards from me, ermines entered our tents when we weren't paying attention, and lemmings attempted to build their nests under our canvas floors.

We worked long hours, especially since the 24/7 daylight allowed us to collect data almost continuously. The perpetual light excited my senses to the extreme. I slept only when completely exhausted and woke up a few hours later to start a new day, wondering if I had missed anything during my sleep. I rejoiced at days filled with new lives: the boisterous charcoal-gray arctic fox cubs playing around their dens; the newly hatched grayish-yellow goslings jostling each other in their nests; the fluffy shorebird chicks sporting speckled plumages that perfectly mimicked the mosses and lichens on the ground. By the middle of my first field season, I had come to realize that arctic fox cubs are so small when they first emerge from their den that they could fit in my cupped hands.

I also learned that the weather in the Arctic was no more predictable than a coin toss and could change at any moment. Some days, I walked peacefully for hours on tundra transformed into golden fields by the low-lying evening sun; other days I returned to camp drenched, cold, and miserable after being caught in a sudden downpour. At times, I battled furiously strong winds that threatened to flatten my tent like a pancake, and had to prop a hiking pole inside to support the tent's roof and stop it from continuously slapping me in the face. Several of my tent flies were ripped to shreds by snowstorms that struck us in the middle of summer. When rainy, windy weather lasted for days, preventing us from cooking or even heating water safely, we ate only bread with peanut butter or cold food directly from the can.

I often didn't shower for weeks. And I quickly learned to sleep as lightly as possible in order to wake up at the first droplets of rain. The sound of rain was an alarm call, prompting me to roll out of my sleeping bag, get dressed as fast as possible, and run to close the fox traps. Unable to shake itself dry, a wet fox can't thermoregulate properly if left in a trap, so to avoid harming any foxes, we suspended trapping as soon as it started to rain.

Three months of fieldwork usually left me worn out and bone weary. Back in my office at the university, however, I was rewarded for my summer's work with glimpses of the arctic foxes' mysterious winter lives as I watched the colorful dots that represented my collared foxes' locations moving day after day on my computer screen. Usually foxes remained on Bylot, but occasionally one would dash away from the island and head north across the frozen land and sea ice to Ellesmere Island, or northeast to Greenland, or southwest to Alaska, astonishing me with their lengthy journeys across the vast Arctic. Some of these travelers settled in new territories hundreds of miles away from their previous homes. A few returned, after months of wandering, to the very same den they had used the previous summer.

As is often the case in science, studying fox movements left me with more questions than answers. What factors influence their decision to leave or stay on Bylot? How do they navigate across vast areas in the dark polar night and return to Bylot without retracing their steps? How do they detect food at a distance and congregate at the remains of seal carcasses left by polar bears on the sea ice?

When summer came, and with it my return to Bylot, I happily reunited with familiar resident foxes and met new ones. Tempête was only one of the amazing foxes I encountered on the island. There was also Flocon, a clever male who split his litter among two or three dens as soon as he spotted our tents nearby (he seemed determined to prevent us from trapping his cubs); Brad, who traveled hundreds of miles during the winter and returned every spring to the same den; Ocarina, who vocalized abundantly as soon as we approached her den; and Heidi, who successfully raised twelve cubs all by herself after her mate was shot by a local hunter.

For my last field season, five years after my first, I stayed on Bylot three weeks longer than usual and captured foxes in late August. Like the first arctic fox I trapped, my last one was also a young female without cubs. As soon as I saw her, I was stunned. She was starting to molt into her winter coat. In May, the spring molt makes foxes look like poodles after a shaving session gone wrong: large patches of long white hair still hang over their short brown summer fur in places, sometimes supplemented by chunks of old fur dangling from their bodies. But during the autumn molt, which I was witnessing for the first time, the winter hair seems to grow in between the summer hairs, and this mix turns the whole coat into a lustrous silvery-white pelt. I brushed the pearlescent silver coat with my fingertips and felt the thickness of this new fur that would allow her to survive the deep freeze of the arctic winter. Like a true spirit of the North, she glistened softly under a late evening sky dimly lit by the last rays of the midnight sun and a timid crescent moon.

As I released her—ear-tagged but without a satellite collar since I didn't have any left—I felt my heart squeeze. I would not know where this beautiful fox went during the winter or whether she would survive. Life in the Arctic is harsh and merciless, and she might face starvation, fail to find a mate, or worse, meet her end in a hunter's trap. Silently, I wished her a good life. That summer, the lemming numbers were low, but since small rodent populations have boom-and-bust cycles, I knew their numbers would go up during the winter and peak the following summer. As she disappeared into the distance, I hoped she would ride that prey wave, turning lemmings into a new generation of feisty arctic fox cubs that would—just as their mother and their ancestors had done for hundreds of millennia—grow strong, put on dense winter coats, and keep roaming the Arctic on their little running paws.

Courtesy of Sandra Lai

Sandra Lai was born in Tahiti, French Polynesia, but her scientific path led her to the cold biomes to which she would one day dedicate her career. Her fascination with the Far North brought her to Canada, where she earned a doctorate in biology from Université du Québec à Rimouski, tracking the movements of arctic foxes from Bylot Island across the Canadian High Arctic. She continued as a postdoctoral researcher, studying arctic hares and polar desert biodiversity at Alert, the northernmost permanently inhabited place in the world.

Based at the University of Oxford's Wildlife Conservation Research Unit (WildCRU), Sandra works with the Ethiopian Wolf Conservation Programme, supporting research and field monitoring of the rarest canid on the planet. Passionate about carnivores, especially canids and small wild cats, she explores how animals move, adapt, and survive in harsh landscapes. Her work continues to take her to remote and fragile ecosystems—from the arctic tundra to high mountains, ancient rainforests, and deserts—where science and conservation meet in the effort to protect species on the edge.

Finding Balance

The Balance

Sarah Winfrey

Women balance career and family every day, but field biologists face some unique challenges in maintaining this equilibrium. My husband and I already had two children when we first put our belongings into storage and moved into a tent in the Idaho wilderness for a summer of fieldwork surveying for flammulated owls. It never occurred to Harley or me to give up fieldwork as the kids grew. To make it work, we traded off duties with each seasonal job—one of us worked in the field while the other homeschooled the kids at our campsite and assisted with the fieldwork whenever possible.

By the time we joined a survey for nesting birds along the Denali Highway in Alaska, we were up to four kids—ages nine, seven, and two-and-a-half years, and a seventeen-month-old toddler. For this particular project, Harley flew to Alaska in early May and began surveying for birds along with four other biologists. The team would study the nesting success of tundra songbirds at different elevations, producing data that would ultimately be used to predict possible effects of climate change on bird populations over much larger areas.

Summer in Alaska is all hustle. With only a few weeks to pair up, build nests, lay and incubate eggs, and raise their young, birds like the white-crowned sparrow are active every minute of the nineteen to twenty-one hours of daily light that Alaska's short summer affords them. Biologists who are trying to monitor these birds would quickly be pushed beyond exhaustion after tracking their subjects for half that time if they didn't have someone who was dedicated to maintaining their field camp. Because the kids and I had already planned to follow Harley to Alaska, it made sense for me to fill that critical camp manager position.

In late May, I gathered our gear for four months of camping in Alaska and puzzle-pieced it together to fit into our Toyota Prius along with all four children. We drove ten hours a day—trekking northwest from the Missouri Ozarks to Dawson's Creek, British Columbia, then winding our way along the ALCAN Highway—and camped each night. At day's end, our older two, Benton and Henry, pitched the tent while I changed diapers, pulled out bedding, set up our camp kitchen, cooked supper, and shuffled though the daily detritus that inevitably settles into a vehicle with child passengers. On our eighth day, we passed through Delta Junction, Alaska, and then drove for a final two hours to our field camp on the Maclaren River, in full view of the glacier with the same name. We pitched tents for our four-month stay—a large one for me, Harley, Ruth, and little James, and a smaller two-person tent nearby for Benton and Henry.

Then it rained. And rained. It rained every day we were in Alaska. I'm not exaggerating. Every. Single. Day. Some days it was just a ten-minute drizzle, others it was a steady shower the entire day. In the rain and cold and wind and rain and sun and damp and rain, I prepared sack lunches for the crew, tightened guy lines on the tents, reminded Benton to close the coolers, pushed away rainwater that had accumulated in the tarp over the kitchen area, found Ruth's missing shoes, cooked supper for nine people in pots designed for four, redirected James from wandering to the river, hung out damp blankets, Band-Aided another of Henry's bumps, and rushed to gather blankets off the line when the rain resumed.

I made a weekly four-hour round-trip to Delta Junction to restock provisions at the grocery store and to wash diapers and mud-caked clothes at the laundromat. On these field trips from the field, I cobbled together ways to lift our dampened spirits. We refilled bags of books at the library for my voracious readers. I purchased a large cheap tent and several beach balls and set up an indoor gym back at camp.

On other days, we walked two miles to collect fresh eggs from the closest neighbor and watch her train her sled dog team. The kids flew two-dollar kites over the tundra. They even opened their own business, collecting wild blueberries and selling them by the cup to tourist buses traveling to Denali National Park. They advertised their wares with "Blueberries 4 Sale" spelled out in rocks on the hillside near our camp.

Once or twice a week, the kids and I would head out with the field crew to lend a hand with the project for half a day. As the biologists searched for new nests and resighted birds marked with identifying colored leg bands, Benton and Henry settled down with binoculars at the edge of the study plot. Tasked as bird lookouts, they radioed their sightings to Harley. Meanwhile, I hoisted James into the kid carrier, and with Ruth in tow picked our way through the scrub, using my GPS to help relocate previously found nests. Once I'd found each nest, I'd record whether the eggs had hatched, how old the nestlings were, or whether the young had fledged. We stayed busy even if we couldn't stay dry.

For a mother, it becomes routine to count heads. Leaving the grocery store—one, two, three, four. At the park after a few hours—one, two, three, four. One morning at about 3 a.m.—Alaskan dawn—I rolled over and automatically counted. One. Just one. Not two. Our toddler, James, was missing from the tent, the zippered flap wide open. My mind went straight to the river—the glacier-fed, rushing-from-spring-melt river—where he had already lost three shoes after creeping too close to its mesmerizing edges. I bolted up from my bed on the ground, shoved my feet into the first shoes I found, and raced to the river's edge yelling my head off. No James. Near panic, I tried to tune in to my field biologist training and scanned for evidence that he had come this way. Thanks to the accursed rain, it was easy to see that the only fresh tracks were my own. I began to dart around the campsite, looking into the older boys' tent, the tarped kitchen area, the parked cars. No James.

Not far in the distance, a grazing moose and her calf precipitated a new anxiety. A cow moose protecting her calf can be one of the most dangerous animals in Alaska, even more dangerous than bears. Bears. I began scanning the hillside for another of our unpredictable neighbors, straining to pick out the shape of a lumbering bear in the soft light. My eyes passed over a small flash of red. I turned nearly full circle before that red registered in my mind.

Snapping my gaze back to that red speck, my eyes locked onto the fleece jacket of a small and delightfully happy little boy. James had found a spot in the blueberry patch and was gorging on the delicious sweetness of that wild bounty. Flooded with relief, I rushed to him and scooped him up. His face and hands were stained purple-blue from the berries, and his smile was infectious. He was so proud that he had remembered his jacket in case it rained.

That day I made an unscheduled visit to town to purchase a carabiner to clip the tent's zipper tabs together and a bell to create an alarm that would hopefully alert me to any future attempted wanderings by our too-independent offspring.

Over the course of twenty-five years, we worked in California, where a cougar stalked the kids around our campsite; in Texas, where I was eight months pregnant and chasing a toddler who wouldn't stop putting cigarette butts in her mouth; in Nevada, where the Mojave Desert afternoon temperatures exceeded 110 degrees for eighteen days straight; and in Idaho, Missouri, Colorado, Florida, Kansas, Nebraska, Pennsylvania, Iowa, Indiana, and Wyoming. For years, though, when I thought about our Alaska work, all I remembered was that it was the hardest field job I'd ever had. I spent that soggy summer watching harried sparrow parents raise their nestlings during Alaska's fleeting summer while juggling the demands of my own offspring and five field biologists. But even though those days often felt interminable, I soon realized that my own hectic time with my kids and my life as a biologist were flying by as quickly as the sparrows' nesting season.

Benton and Henry have now graduated from college with environmental degrees, and Ruth and James aren't far behind. All are committed to making a difference in the world. For years I struggled to maintain a balance between nurturing my kids as we raised them in the wild and fulfilling my passion for working as a field biologist. Watching these young adults fledge into the world one by one, I am beginning to contemplate a life less dependent on maintaining that precarious equilibrium. The struggles of that rain-soaked Alaska summer and all the rest of it have acquired a rosy glow. In retrospect, it all went far too fast. And all of it—every hard moment of it—was so very worth it.

Harley Winfrey

Sarah Winfrey grew up as a Girl Scout in Missouri and knew she wanted to be a wildlife biologist by the time she was ten. With the help of a college ornithology class, she recruited her fiancé from political science into wildlife biology. They began hopping around the country doing temporary field jobs during their summer breaks, and their first child was born one month after graduation. The new recruit didn't slow their travels, and over the course of their careers studying frogs, birds, bats, insects, and more, they added three more kids to the field team, trading places as the field biologist and "stay-at-home" parent while living in tents, trailers, apartments, and field houses.

Sarah works in the Mojave Desert monitoring birds and tortoises. She and her husband and their two youngest children still move every few years following biology work, while their two oldest children are pursuing careers as a forester and an environmental engineer.

Walk the Desert Alone

Andrea Hazelton

In loving memory of Larry Webb, who was like a father to me for many years

"Are you sure you don't want a gun?"

My father-in-law, Larry, who never minded having the same conversation many times, brought up his new favorite topic. He was a Vietnam veteran and also excessively protective of his family.

"I don't think I'm allowed to carry a gun."

This was sort of a lie. Fresh out of college, I'd recently started my first "grown-up" job—as the botanist for the Fish and Wildlife Department of a Native American tribal government in the southwestern US. During my first day on the job, one of my bosses had told me I could carry a gun in my work truck as long as I didn't tell anyone about it. He had been very specific, however, in making me understand that I was not permitted to transport any dogs in my work vehicle—not even for personal protection.

"They can't take away your right to have a gun!" Larry was never one to give up on a topic when he felt passionate about it. "How are you supposed to protect yourself? You just say the word, and I'll buy you a gun."

"I don't need you to buy me a gun. I already have a gun." Born and raised in Arizona, I was an ecologist married to a big-game hunter. Unlike many of my friends from back East, I had developed a nuanced view of gun ownership. I wasn't terribly interested in having or using one, but I could appreciate their utility. After all, I liked eating meat, and guns were awfully helpful for getting meat from the hoof to my dinner plate.

A year or two earlier, my own father had gifted me the one gun he owned, a .22 caliber rifle. I didn't know much about guns, but Dad had explained to me that this gun shot .22 long rounds and was semiautomatic, which meant that after I fired a shot, I didn't have to pull a lever or do anything else to pop out the empty bullet casing. I could just pull the trigger again to shoot the next bullet.

"It's for plinking," Dad had said. I wasn't too sure what "plinking" was, but I thought maybe it was a redneck word for shooting at an old TV.

Larry was not deterred by being reminded of my status as a gun owner. "No, you need a gun that will fit under the seat of your truck! I'll get you one."

"I'm really not interested, okay?"

At this point, I went to the bathroom in order to end the conversation. I didn't want to get to the inevitable point where I would demand to know exactly what I needed protection *from*, and he would respond by running through an absurd list of supposedly life-threatening wildlife species and finally point out that after all, I might even be attacked by a man.

The issue at hand was that for my new job, I was expected to do fieldwork alone—largely a new experience. In graduate school, I'd done plenty of fieldwork with other students and even spent occasional days working on my own. But I had never camped alone.

The thought of it made me nervous, even though I would never have admitted that to my overprotective father-in-law. Accordingly, I managed to arrange volunteer assistants for my first three overnight trips to the field. The first trip was an autumn backpacking excursion into Tsegi Canyon, a gloriously dramatic canyon cut into salmon-colored sandstone, with nooks and crannies and hanging gardens around every corner. I brought my younger brother, the photographer, who was excited to accompany me. We looked for endangered sedges as we explored the canyon. He taught me how to use my new work-issued camera with its fancy macro lens for taking close-up pictures of flowers. And most memorably, we spent one damp and malodorous night wrapped up under a single tarp, our only shelter from an unanticipated rainstorm.

The second trip, later that autumn, was to nearby Potato Canyon, which had many of the same aesthetic qualities as Tsegi Canyon but was more remote. I brought my friend Charlotte, the park ranger. She provided moral support as I climbed up to narrow ledges, lay on my belly, and eased out to even more precarious locations, where I used a magnifying hand lens to identify sedges. We made coffee using her French press and my brand new super-compact backpacking stove (hooray for grown-up paychecks!), and learned together how to use a topographic map in conjunction with a GPS unit and thirty-year-old field notes to figure out where to look for plants (while managing to get a bit lost a couple of times).

As we worked our way up a side canyon on day three, we faced an unfortunate choice. We could either crawl through a dense thicket of spiny shrubs or scramble down a nearly vertical dirt bank and walk up the stream channel. Not being too keen on getting spines in our faces, we opted for the stream. After sliding our way down the dirt bank, we reached the sandy stream bottom. And then kept sinking. This was not my first experience with quicksand, and I knew how to react. Don't stop. Keep moving fast and be light on your feet—so it can't eat your shoes—until you reach dry ground. I did just that and then turned around to check on my friend. Charlotte, unfortunately, had had no prior experience with quicksand, and she had not moved at all.

"It's okay!" she exclaimed. "I just saw a TV show about this!" She still hadn't moved.

At least she wasn't breaking rule number one: *don't panic*. But she was carrying my brand new work-issued camera with its fancy macro lens.

"Charlotte, the camera!" I knew my friend was going to be fine, but I feared that the camera would fare worse than she would if it got sand grains in its crevices. Charlotte's eyes got wide, and she looked around at her options before gently tossing the camera, in its padded case, onto stable ground.

Moments later, Charlotte still hadn't moved. I edged closer to her, keeping to relatively dry ground just above the quicksand in the channel bottom. "Do you need a hand?"

"No, I've got this." She took off her backpack, tossed it to where she'd thrown the camera, and then pitched herself facedown onto the stream's dirt bank. Belly down, arms out and flat on the ground, with her face in the mud, she wiggled her way upslope and extracted her feet from the quicksand. Fortunately, her hiking boots emerged along with her feet.

I wasn't sure if it was too soon to laugh, so I kept it to a grin until I saw the wide smile on Charlotte's mud-smeared face. After sharing a laugh, we sat right down on that dirt bank and ate our tuna, crackers, cheese sticks, and apples. We then weighed our options: keep working up the side canyon, survey plants in a different location, or head back to the truck. Charlotte is reasonably tough and has no tendencies toward drama, but that day she wasn't afraid to admit she was ready to go home. So we hiked back to the truck and did just that.

My third overnight trip came the following spring. My objective this time was to contribute to a long-term monitoring program for an endangered milkvetch species. The plant was known from only one location, near the mouth of a tributary to Lake Powell. I had read that the plant in question was being eaten by wild burros, which I found difficult to believe because many closely related plant species are inedible and even poisonous. So I had called a wildlife biologist I'd met at a professional meeting—an expert in horses and burros—and invited him to check out the situation with me. Martin had been happy to oblige.

He met me in Gallup, New Mexico, and hopped into my truck, and I then drove us toward Lake Powell and down a canyon. The long drive to the site started with four hours on the highway and ended with several unexpected hours of using four-wheel drive to navigate down a mostly dry wash at the bottom of the canyon.

Martin showed me how to tell burro tracks from horse tracks, how to differentiate burro and horse droppings, and how to tell if a burro was feral based on how closely it allowed you to approach before it ran away. According to his test, these burros were feral. He helped me design a new way to document the pressure from grazing burros and figure out whether the endangered plant population was in decline because of their feeding habits.

We camped one night in that canyon, and as Martin built a fire from cottonwood twigs, it occurred to me that in a way, camping with this near-stranger was more dangerous than camping alone would have been. We were many miles and several hours' drive from another human being. I had to trust that this man, whom I barely knew, was a good person. I thought of Larry and the sentence I did my best to avoid hearing him say. *After all, I might even be attacked by a man.*

Everything turned out fine, and I never had cause to believe that Martin was anything but a good person. But that was the night I realized that when I slept in the wilderness, I no longer needed to use the buddy system.

With a few exceptions, I did the rest of my fieldwork alone for the next nine years. I will admit that I was a bit uncomfortable in the beginning. At age twenty-six I was, like many people, unaccustomed to solitude. I had moved from my family's house to college dorms, lived with friends in a small apartment after that, and then gotten married and moved in with my garrulous, extroverted, music-loving husband. I had been living with him ever since, and there was rarely silence in my life. That is, until I began doing fieldwork solo.

My mild discomfort manifested itself primarily in a behavior I came to equate with drunk texting. This usually happened during the first few nights of my first several trips every year as I readjusted to the solitude. I would finish my day's work, return to the truck, and make dinner. After eating, I'd clean my dishes and put my kitchen things away.

With my responsibilities completed, what was left of the evening would stare me in the face, and so I would begin to text. After clambering onto my truck or climbing nearby hills to try to get phone service, I would send out a steady stream of texts—to my husband (he might be worried about me!) and to a long list of friends—to which I often received no responses. But that didn't dissuade me from texting and texting, as if just the hope of getting a response was enough to connect me to the rest of the world.

The sad thing is, I had little to say. When you are working alone in the field, things rarely happen that other people might find noteworthy. There was the howl of the wind in my ears and the echo chamber inside my own brain. I might have seen a red-tailed hawk or a golden eagle or several kangaroo rat middens or more than a hundred endangered cacti. Maybe I'd almost gotten lost when trying to find the field site. Maybe I had worked faster than expected and would be able to drive home early, or maybe I had been foiled and would need to stay an extra day. Had someone else been with me, we might have talked about these things and maybe even embellished them when sharing them with others. But when these tiny events had only a single lonesome witness—me—they somehow lost their import.

With time and experience, my compulsive texting waned, but I did develop a preference for campsites that had cell phone service. It seemed sensible to maintain a daily link—or at least the possibility of a link—to humanity.

Unfortunately, as my obsessive texting waned, another, less socially acceptable behavior emerged. I began talking to myself—mostly for planning purposes. With nobody there to hear me, I would explain to the howling wind exactly which path I was going to follow to complete a survey, how long I expected it to take, how much daylight would be left when I finished, and which campsites I might be able to reach before dark. Sometimes I did math out loud or worked through my to-do list for the week. Talking through a problem often helps me to focus on it, and it turns out that having an audience is not important at all. Several years later when I began bringing my dog into the field, this habit grew worse, but it became acceptable. After all, I was no longer talking to myself—I was talking to my dog.

One of the more useful habits I developed while working alone was exercising basic caution. This habit manifested itself in a variety of ways, from what I carried in my backpack to the emergency gear I kept in my truck to where I was willing to drive or hike. I needed to be fully self-sufficient, and if I injured myself or stranded my truck, I might need to walk many miles before I could call for help. Caution also guided my campsite selection.

For women who camp solo, there is an art to choosing a campsite. I make this generalization for two reasons: first, because I have discussed the topic with the few women I know who regularly camp alone, and they all admit to employing similar tactics; and second, because it has become clear to me that my male friends who do similar fieldwork put almost no thought into choosing their campsites. A male colleague once told me about a time he parked on the edge of a dirt road in the middle of the night and set up his cot next to his truck. In the morning he woke to the sound of dogs barking, opened his eyes, and discovered he'd camped in someone's front yard.

When I camp alone, I make a plan well before dark. I decide early in the day where I intend to camp that night and what route I will take to get there. My primary goal is to be out of sight—and thus out of mind—of anyone. It helps if I'm familiar with the area and know which roads are likely to be well traveled, particularly during the evening hours. I make sure my camp is not visible from such roads, if possible. Dense pinyon pine forest is my favorite camping habitat because I can turn onto a two-track woodcutting road, follow it for fifty yards, and be completely hidden from view.

As a backup safety measure, I try to camp incognito. When I worked for the tribe, I never once set up a tent. I always slept in the bed of my truck with the tailgate up. I kept all my gear inside the cab or next to me in the truck bed. If anyone did happen to drive past, they would only see a parked truck, not a camp. I disabled the dome light in the cab so it no longer turned on automatically when I opened or closed a door. I used a headlamp, never a lantern, and only when I needed it to see something in particular.

These personal guidelines have never let me down. After I left my job with the tribe and began working in national forests more often, I started allowing myself the luxury of a tent, as long as I was camping in an area where campers would normally be found. But I still tried to camp out of sight. In nine years of solo camping, I've never been approached, let alone harassed, after dark.

For me, to work and camp alone is to feel free. I work on my own schedule—even though that schedule tends to be a grueling dawn-to-dusk one. I don't have to wait for anyone else to pack their backpack before starting work in the morning, and nobody is cranky because they want to put the breakfast things away when I haven't eaten yet. I make my own safety decisions without having to worry about other peoples' phobias or knee pain. And nobody tries to converse with me before I've had my coffee.

In the evening when I get back to my secluded camp, I can crack open a beer, heat up some water on my camp stove, stand naked next to my tailgate, and bathe. Then, if the weather is warm enough and my camp is secluded enough, I can sit naked on my tailgate and watch the sunset.

In the seven years after Larry gave up trying to convince me to carry a gun, some major life changes dulled the naive shine I had worn as a fresh-faced college graduate. After several job changes, I began working independently as a biological consultant, taking many small field jobs each year. I went through a divorce and a couple of other breakups. The last of those breakups was the catalyst that finally led me to get my .22 out from under my bed.

Eric and I had dated casually for a couple of years. We'd never been serious or made plans to be together long term. For about six months, while we'd each been traveling for work, we had seen very little of each other and had made little effort to connect when we were between jobs. For me, the relationship had run its course. Then, while working a springtime job in the Grand Canyon, I met another man.

I had no phone service where I worked in the canyon, but when I got home from that job, I called Eric and gently told him that our relationship was over, that I had met someone else. I was shocked when he began screaming at me over the phone. I hung up. That night he came to my house, something he hadn't done in months. He pounded on the front door and then the walls, circling the house and bellowing insults. I called the police. He disappeared into the night as soon as the police car showed up. The policeman waited and kept watch while I packed an overnight bag, climbed into my truck, and drove to my new boyfriend's house, no longer feeling safe in my own home. In the morning, a close friend called and told me that Eric had showed up at her house in the middle of the night looking for me. Twenty minutes later, my mother rang to let me know that Eric had telephoned my father and called me a string of insulting names.

Feeling shaken, I phoned a few friends and asked them to meet me at my house. While they stood watch outside, I packed my camping gear and enough clothing to be away from my home indefinitely. Then I locked the door and left.

I knew I had to leave the small town where Eric and I lived. My roommates, who were also field biologists, were working out of state throughout that spring and summer, and I didn't feel safe alone in my house. I could not go to my parents' house, two hours away, because he would surely track me down there. So I drove east, to New Mexico, where I had friends Eric had never met. I stayed with friends in Gallup for a few nights and then camped with my new boyfriend for a week, until I had to travel to my next job.

For six months I lived in my truck, going from job to job and staying with friends in between. I paid rent on my house every month, but I didn't dare go back there. Eric called at least once a day and often in the middle of the night. I never answered. His voice messages raged one day and apologized the next. He sent me insulting texts, stole packages that were mailed to my house, and threatened to harm me.

For the rest of that spring, summer, and autumn, the field was my safe space, where he couldn't find me. And when I was alone in the field, I felt not only physically safe but emotionally safe as well. I was free to work through the pain, fear, and humiliation, and to wrestle with my inability to comprehend how this could possibly be happening to me. I walked for hours with tears running down my face, and nobody saw me. I screamed at Eric and at the universe, and nobody heard.

I needed more than anything else to be alone in the field. It was a struggle for me to work with anyone else, even close friends. I cried too frequently for others to be comfortable around me. On one occasion, when my boss got frustrated about some job logistics, cursed, and threw some paperwork around, I leaped from our parked vehicle and cowered nearby, involuntarily shaking.

I suppose I had PTSD, although I never talked to a counselor about it. I just worked, walking back and forth and back and forth, surveying my vegetation transects. Sometimes I walked the open desert, in straight lines among sparse saguaro cacti and silver bursage bushes, cutting obliquely across wide sandy washes dotted with ironwood trees and wild heliotrope. Sometimes I walked the mountains, going straight up one side and down the other side, moving thirty meters from my previous path, then going back up and back down, over and over, looking for endangered plants and mapping the places where I found noxious weeds. I sat down at 11 a.m., ate my sandwich, then kept walking. I ate my snack at 3 p.m. and kept walking. I climbed over fallen logs, squeezed through thickets of spiny locust trees, huddled under overhangs during lightning storms, and slogged through mud and charcoal from burned forests, all the while consumed by my pain and fear and confusion.

I practiced my breathing. In-two-three-four. Hold-two-three-four-five-six-seven. Out-two-three-four-five-six-seven-eight. Until eventually, the fear and the pain began to recede.

Finally, my field season came to an end and I had nowhere to go but home. My roommates had returned, so at least I wouldn't be alone in my house. In the first weeks after my return, I saw Eric around town—more than would be expected by chance—and each time, it shook me. Worse, every week or two he would stand in the street outside my house and scream his poor opinion of me so that all my neighbors could hear. I filed for a restraining order, but law enforcement apparently had trouble finding Eric to serve him. Before long, my roommates headed off on other adventures and I was alone in my house again. That's when I got out my gun.

My .22 rifle is about three feet long, with a simple, vintage look and a wooden stock that's a warm red color. It had been so many years since I had shot my gun that I wasn't sure if I even owned any ammunition for it. My first night alone in my house, I pulled it out from under my bed, took it out of its case, and leaned it up against my bedside table. I also made sure my phone was sitting next to it. I had a plan. If Eric showed up and managed to get into my house, I would dial 911, set down the phone, pick up the rifle, and point it at him. If that didn't stop him, I'd clock him over the head with it. This, I thought, was better than having no plan at all.

That gun leaned against my bedside table for more than a year.

Finally, Eric stopped showing up outside my house and yelling at me from the street. When I walked out of yoga classes, I no longer spotted him leaning against the building across the street, pretending to look at his phone.

And so, finally, I put the gun back in its case and stowed it under my bed. Now I live in a different house in a different neighborhood and drive a different car. I'm harder to find, and it's been a couple of years since he last approached me. My rifle lives in its case under my bed. But I still watch my back when I'm in civilization, and I always lock my doors. While my current job rarely provides opportunities to do fieldwork alone, I go for long hikes with my dog on the weekends. Sometimes we even camp out in the desert, just the two of us.

To this day, I have never once been tempted to bring a gun.

Andrea Hazelton is a field botanist who works throughout the southwestern US. She was born and raised in Phoenix, Arizona, and attended Arizona State University. She began as an English literature major but quickly fell in love with ecology and botany while taking introductory courses to fulfill the minimum science requirements, and changed her major soon after. After receiv-

Chad Smith

ing her bachelor's in plant biology, she stayed at Arizona State to earn a master's degree, working with her favorite ecology professor on a study of riparian plant ecology along desert rivers.

After graduating, Andrea became involved in endangered plant conservation, first as the botanist for the Navajo Natural Heritage Program and later as an environmental consultant specializing in rare plants of the desert Southwest. During the winter, she lives with two kind, supportive housemates and two adorable dogs. During the field season, she spends most of her time on the road, living out of her four-wheel-drive pickup truck.

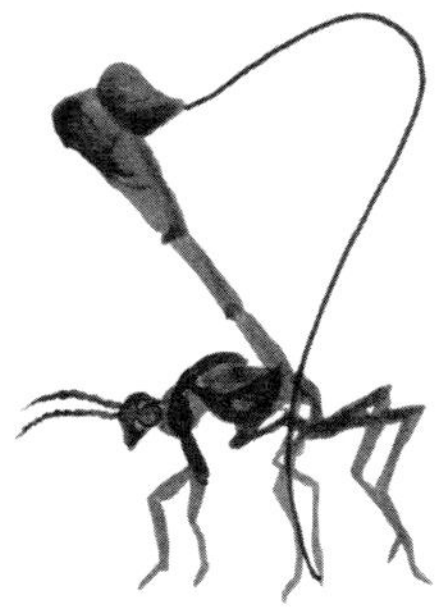

Fig Lessons

Marlene Farrell

Bookish learning, number crunching, staring at a computer screen late into the night—that was a necessary phase that I survived. At long last, I received my piece of paper—my ticket to finally be the one tramping through the bush, binoculars around my neck and notepad in hand. My first opportunity? At twenty-two, I was offered a job assisting a Smithsonian Tropical Research Institute biologist on Barro Colorado Island (BCI) in Panama. I couldn't wait to see the island's biological wonders for myself: leafcutter and army ants, canopy-dwelling orchids, spider monkeys, agoutis, tarantulas, toucans, fruit-eating bats, and more.

Even so, I was unprepared for how different it all was. Approaching BCI by passenger ferry for the first time, I saw a dark green mound—a towering and impenetrable forest—rising out of the Panama Canal. I imagined the forest to be full of closely guarded secrets. BCI became an island only in 1914 when the Chagres River was dammed during construction of the canal. The resulting Gatun Lake covered the existing tropical forest, surrounding and isolating hilltops that became islands. Since the 1930s, scientists from around the world have come to BCI—the largest island at thirty-two hundred acres—to study how artificial isolation has affected its flora and fauna, and to conduct long-term research on changes in tropical forests over time.

As we neared the island, I could see the rooftops of a smattering of buildings—research labs and dwellings. Since the island hosted several dozen scientists and graduate students at any one time, it had the stamp of human presence, but if the station were to be abandoned, it clearly wouldn't take long for the forest to reclaim the space, with vines twining up the balconies and monkeys hopping from roof to roof.

Those first weeks, when I walked through the forest, sticking to the narrow winding paths and endless steep steps made of concrete blocks, I felt claustrophobic. Fecund, verdant vegetation arched over me and leaned in, growing in front of my eyes. The air was thick with moisture. Light trickled down through small gaps in the canopy, never pooling on the forest floor. Sounds of forest life hit my ears: hordes of tiny feet scrabbling and crawling through the duff, up the tree trunks, and along the leaves; the flutter of wingbeats and snippets of song cascading from invisible heights.

This was to be my new home for many months.

My boss and his team, made up of a graduate student and a few research assistants, studied how figs and fig wasps have evolved together, a relationship that has been millions of years in the making. I'd learned about this intertwined evolution between figs and fig wasps in college, but it seemed like a fanciful tale until another project assistant, Eileen, and I were motoring around the island's perimeter in a small boat, knocking figs out of the trees to bring back to the research station's lab for dissection.

We enjoyed being on the water, spotting one of our flagged trees, puttering up to it, and squeezing the boat among the hanging branches. Then with long loppers one of us would snip a few of the branches, covered in small green globes, so that they fell into the boat, until we'd collected enough for our research.

Figs, these unremarkable-looking green spheres, are inverted flowers rather than actual fruits. Clusters of tiny flowers are contained in a fleshy receptacle (the fig) that can only be accessed by a specially adapted, tunneling pollinator (the fig wasp). Hundreds of fig species are found in tropical areas all over the world, and each is associated with at least one unique wasp species. As they grow, the figs become self-contained ecosystems. While inside the fig, the female wasp pollinates the flowers and lays her eggs. The baby wasps hatch, feed, grow, and even mate within the fig, where they're joined by interlopers—the progeny of nonpollinating wasp species that have parasitized the fig, and parasitic nematodes that have hitched a ride into the fig concealed in the body of a wasp.

Encircled by the waters of the Panama Canal, BCI, like the figs, hosted a somewhat self-contained community—this one consisting of a hierarchical network of scientists, graduate students, field and lab technicians, and support staff. Most of us stayed long enough to become a pseudo-family, eating beans and rice, papaya, and fried bananas together in the cafeteria, and occasionally sipping Panamanian lagers on the balcony. Our numbers shifted slightly over time, as someone returned to their respective university or a new researcher arrived.

When I wasn't in the boat or engaged in other field tasks, I worked in the research station's lab, where I would slice a fig in half, peek at its inner mayhem, and count its inhabitants. As I looked through the microscope, I could spot the expired, crumpled adult fig wasp and her white oblong eggs, each deposited into a fig flower ovary, surrounded by food to fuel its development. I could see intact seeds as well, growing thanks to the pollen brought in by the now-deceased fig wasp; the fig and the fig wasp would not have coevolved as "friends," mutually benefiting by their association, unless both had been able to successfully produce offspring.

When a fig was soft to the touch, like a ripe avocado, I could open it and watch the magnified eggs hatching into writhing fig-wasp larvae. Within a couple of weeks, these larvae would transform into adults, mate (including with their siblings), collect the fig's pollen from intact flowers, and either die inside (males) or crawl through freshly made tunnels and fly off to start the cycle again (females).

The story of figs and fig wasps could be viewed as cooperation, since both organisms benefit. Or it could be viewed as an arms race, with each species evolving defenses against the other. When the female fig wasp initially squirms through a tiny specialized opening into the fig to lay her eggs, her wings usually break off because of the tightness and roughness of the opening, so she can't escape. Hence her chosen nursery often serves as her tomb. The fig pays a price, too. For every egg the female wasp lays inside the fig, there's one less flower that will produce a seed. Later, the wasp larvae will devour even more seeds. Still, the ever-evolving association between figs and fig wasps enables the wasp offspring to thrive—because they have been well protected inside the fig—and simultaneously ensures that the fig's pollen is shuttled to other trees.

After the next generation of fig wasps has escaped, the ripe figs entice mammals and birds, which feed on the fruit, then excrete the seeds, dispersing them to new germination sites.

Looking inside the fig, I also investigated the story's minor characters, which added to the drama. These included several species of nonpollinator wasps that used the fig as a nursery without giving anything in return. Wielding the long ovipositor at the end of her abdomen like a needle, each parasitic wasp had punctured the fig and transferred her eggs into its flesh. The larvae that developed from these eggs stole resources from the fig and sometimes killed the fig wasp larvae.

Finally, I learned to make the microscope's light gleam off the tiny wriggling strands of vermicelli-like nematodes—microscopically small roundworms—that initially infiltrated the fig by hiding inside the fig wasp. After escaping the wasp's body, the nematodes poached nutrients from both the fig and the expired fig wasp, raising another generation of worms to infect the developing daughter wasps and begin their cycle anew.

Did this cozy but brutal community disturb or fascinate me? I wasn't sure what to think.

The society on BCI, with its unwritten rules, also puzzled me at first. The professors and scientists were older, often married. They stayed at the station for a few weeks to set up their research and then left it in the capable hands of graduate students and assistants, while they returned to their university to teach, analyze data, and write papers for publication. The rest of us were single twenty-somethings. This crowd was diligent, studious, and nerdy, but working ten-hour days six or seven days a week still left time to indulge in the freedom of a temporary situation that was disconnected from home.

Swayed by active hormones, the younger crowd indulged in courtship rituals that mirrored some of those we observed in the surrounding wildlife, our romantic or lustful posturing adorning us like so many elaborate feathers. At dinner my very first night, a grinning graduate student offered to take me up into the canopy using his ropes and ascenders. I accepted the chance to view the world from his high vantage but successfully deflected further attentions until another boat unloaded more-willing partners for him. His stratagem was fairly innocuous, but I observed others whose efforts were a little more wolfish, and some heartache inevitably ensued. Like our wild relatives, we must weigh the costs and benefits of a relationship that may be fleeting or serve one party more than the other. In this unique environment, I chose to model myself after some of BCI's independent women, whose ambitions and goals helped them chart a path that commanded respect and allowed them to take advantage of all that the island's life offered.

We were an international jumble, drawn to this famous research station. BCI is one of the world's premier study sites for cataloguing and understanding biodiversity. It contains a 123-acre plot established in 1982 in which nearly every single tree has been documented and monitored over time. That's about 240,000 individual trees representing three hundred species! I learned to identify some of the key species, but most of that fantastical diversity remained unknown to me. I never grew fully accustomed to the forest's cluttered and obscuring lushness. Being on the water, where I could feel the sun on my skin, gave me moments of relief from the enclosing forest. I became a swimmer, along with some of my colleagues. We swam almost daily, even with crocodiles lurking along the island's edges.

I had grown up with structure, rules, and tidiness, and sometimes felt tipped off balance by the unpredictability of things on BCI. But I learned to be more comfortable with discomfort and uncertainties. Mostly it was small irritants like a rash of chigger bites under my waistband, living without air conditioning (except in the computer lab), and leaving the island only infrequently to go for a real run or to drive into the city for a movie and a milkshake. Bouts of social discomfort came, too, when I missed friends and family from home or I wanted space away from the demands of the ever-present community. I learned to notice order lurking in the disorder. Once settled in, we human inhabitants of BCI, like the inhabitants inside a fig, could find an equilibrium between self-interest and cooperation—or even interdependence.

Eileen and I worked side by side. While hiking to study plots or entering data, we spilled the viscera of our lives, and I felt a sisterly bond with her. She made me laugh so hard that I didn't mind the long hours or notice the grime that covered me. On the mainland, she drove, singing bits of tunes a cappella. I admired her; I was a terrible singer and a terrible stick-shift driver. On a rare free night, we'd head into Panama City, sometimes with others, to our favorite trattoria, eating bruschetta and sipping red wine under palm trees lit with fairy lights.

Panama's colorful society and extravagant wilderness fed my imagination. Grunt work mixed with new sensory experiences became routine. Equipped with curiosity and humor, I navigated the challenges of fieldwork and being in a foreign place so far from home.

I got used to sweat as an extension of my wardrobe. I dabbled in fancy Panamanian cigars until I admitted that I thought they stank. I experimented with who I was, beyond the girl with good grades and intense focus. It turns out I could toss a mean Frisbee and learn to love Bob Dylan and even try to dance the merengue, letting my hips move of their own accord. My newfound bravery was fueled by an overwhelming gratitude for my constant contact with the teeming beauty of the tropics. Such encounters transformed me.

When I climbed the concrete stairs in the dark, glowing arachnid eyes peered back at me. It made me smile because I've always loved spiders, and I was thrilled to see a tarantula whenever one ventured out of its lair. One evening as I jogged up to a landing near the dormitories, I was surprised by something smacking into my forehead; the bat was even more surprised than I was. A graduate student friend took me into the forest to see another kind of bat. There it was, his study species, a petite ball of sleeping brown fuzz and folded leather tucked inside the shelter of a tented leaf, which it had shaped into a cone with its own dexterous little claws.

Another time, I stumbled across a throbbing beachball-sized mass made up of hundreds of thousands of interlocked army ants clinging to each other inside a tree cavity, collectively forming the nursery in which they would raise their offspring. I feared getting close, even though they were too busy to attack me. Their intense protection of the laying queen and her eggs in the center of the mass would later shift, when the ants disbanded and dropped back to the forest floor to consume any insects (and even small vertebrates) in their path, while still transporting the queen and larvae during their nomadic travels.

The worldwide ubiquity of fig plants is owed to their coevolution with particular fig wasp species, wrought over millennia, but also to the plants' fecundity. Every two months a fig plant produces a new batch of figs. Because the fruit is available year-round, many wildlife species rely on figs as a key food source.

In the same way that figs and their wasp communities repeat their ripening cycles, BCI cycles through scientists, graduate students, and assistants year after year. Every completed study yields results and suggests further questions to be investigated. Given BCI's rich environment, the discoveries will never cease. Dogged researchers continue to wrest answers to their questions in exchange for an offering of sweat and a small blood donation to the resident insects. Their sacrifices aren't to the scale of the female fig wasps, which rarely leave their chosen fig. But the efforts of tropical biologists and their assistants provide an understanding of our natural world that is disseminated like so many baby wasps flying off in search of future fruits, perpetuating a new generation of inquiry and discovery.

I left BCI more than two decades ago, but I'll always remember the burst of shimmering blue when a giant morpho butterfly crossed my path. And the haunting chorus of a troop of howler monkeys, which heralded every dawn and dusk. And the tart but sweet taste of guanabana. Exuberant-yet-enigmatic Barro Colorado Island, the exquisite intricacy of figs, the lustrous delicacy of fig wasps, and the kinship I found in that unique place helped shape who I am today—someone who is better able to handle the gorgeous messiness of life and who will pass my awe and gratitude for it on to my own children.

Kevin Farrell

Marlene Farrell grew up in Virginia, where the ants and praying mantises in her yard fascinated her as much as trips to the National Zoo. This interest led her to earn bachelor's and master's degrees with a focus on behavioral ecology. She conducted fieldwork abroad as well as in Washington State, her home for more than twenty-five years. While raising a family, she detoured into freelance writing, coaching outdoor sports, and nonprofit work, though curiosity about and care for the environment are ever close to her heart.

She serves as executive director of the nonprofit Sustainable NCW, educating, inspiring, and collaborating to help reduce her community's carbon footprint through waste reduction, clean energy, and reduced vehicle miles traveled. Marlene loves glimpsing her wild neighbors while she runs, backpacks, Nordic skis, and sails with friends and family.

Trapped

Marie I. Tosa

We waited and waited. Thirty minutes passed. Then an hour. Then two. No improvement. Still.

The rough and austere landscapes in old westerns provide the perfect backdrop for the resilient and uncompromising cowboy to demonstrate his mettle. So too does the natural world provide the setting for stories of success and accomplishment told by hunters, adventurers, fieldworkers, and biologists. The storyline has a satisfying structure: someone sets out to do what is supposed to be a straightforward task, an unexpected set of surprises tilts the task toward calamity, and then the protagonist overcomes all adversity, conquers all obstacles, and comes home a hero—with an epic story to boot! But that night, mired in a story that had careened into the disastrous, I found myself struggling with disquieting questions that offered no guarantees of a heroic outcome. *Why was she so still? Would she wake up? Would she get up and wobble off? Would she escape this nightmare? Would I?*

We had undertaken what we thought was a valiant mission: a scientific study to help understand the social behavior of white-tailed deer and how management actions could limit the spread of chronic wasting disease, a devastating illness affecting deer and other cervids. To that end, we were capturing deer in southern Illinois and fitting them with GPS collars to track their movements after capture.

On this day, we had tranquilized a doe and her two fawns and fitted them with collars. The doe and one of the fawns had woken up and trotted away without incident. But one remained, prone and motionless. My colleague and I had decided that the rest of the crew could now go home. "We can do this," we had said. The task, after all, was so simple. We just needed to watch over the fawn until she recovered from the tranquilizers. Two hours earlier, when her family left, she had awakened and attempted to rise, but her back legs wouldn't support her. Noticing my panic as I watched her falter, my more experienced colleague had reassured me, "Don't worry, this isn't unusual. Tranquilizers wear off last in the glutes—the biggest muscles—so, often the back legs take some time to wake up." But now, as the truck with the rest of the crew drove off, my panic was replaced by dread.

In theory, trapping deer is a straightforward process that involves following well-established steps: scout for deer, train them to eat from bait piles that we replenish daily, set up a net that drops by remote trigger, then watch from a distant spot and press the trigger to drop the net when the deer mosey in for some delicious golden kernels of corn. Once they are in the net, we physically restrain them and inject them with tranquilizing drugs. Finally, we fit them with a GPS collar and wait for them to wake up and trot off. All in all, from capture to release, the process should take less than an hour.

The previous night's successful capture and release had boosted my confidence. *Trapping deer is easy*, I had thought. *You just have to be prepared.* But the question that tormented me now was: *Be prepared for what?*

Education can be viewed as learning what to be prepared for and how to handle whatever comes. I knew how to collect data. I knew the different ways to capture deer for research: clover traps, rocket nets, drop nets, and dart guns. I knew how to prepare a trapping location, how to set up each kind of net, and how to shoot a dart gun. I knew the protocols and uses of immobilizing drugs—Telazol sedates and Xylazine relaxes the muscles—and the precise doses to administer according to the size of the animal. I knew how to prepare a capture kit and how to program GPS collars. I could clearly communicate to my crew how all these methods and mechanisms worked. I knew the postcapture celebratory rituals, too: laughing, eating, drinking beer, relaxing. I even knew the rituals that followed those rituals: chatting around fires, at bars, and at conferences about overcoming odds, about persevering—just a handful of cowboys gossiping about the day's events.

But what I didn't know and wasn't prepared for was watching the agonizingly slow deterioration of life. Here, there was no ritual. There was no rule—except, as I would learn later, to try to suppress despair, to try to forget. And there would be nothing—no tale of triumph—to proudly share later; the fawn was clearly deteriorating. No longer trying to stand, she lay with her head down, her eyes closed. We had hit the six-hour mark. We had waited far longer than we should have, clinging futilely to the hope that she would miraculously trot away. Out of desperation, we trudged quietly over to her and tried lifting her up onto her feet. She immediately collapsed to the ground. We knew then that nothing could be done for her. We knew then what we would have to do.

The necropsy didn't happen for weeks. It would have taken a well-trained and committed kind of callousness for me to perform a necropsy any sooner after a death I felt responsible for. I didn't have that kind of training. And it took some time for me to feel even a degree of emotional detachment from the fawn, given what had befallen her. Eventually, I worked up the resolve to look at the poor fawn again.

This time, when I set eyes on her, she was a cold, lifeless lump. An unfortunate mystery to unpack and solve. The necropsy revealed that the fawn had been in good health, was a good weight with good fat reserves, and showed no signs of disease. But her spine was broken, which paralyzed her hind legs. We had suspected that her lack of mobility was related to the capture event, and this was confirmed by the contusion we found along her vertebrae. This was not an injury that this animal would have recovered from. Ending her suffering had been the right call. For that, I felt a slight bit of relief.

But the loss of life—unintentional though it was—was not something I could get over easily. I grieved. And I racked my brain searching for ways that this incident could have been avoided. There were three possible explanations for the fawn's broken back: the heavy weights attached to the net to make it drop quickly might have landed squarely on her back; or once caught under the net, she might have thrashed so violently trying to escape that she damaged herself; or our weight might have inadvertently crushed her spine when we physically restrained her to administer the tranquilizer. The last explanation was unlikely, but it was impossible to determine which of the other two possibilities was responsible for her subsequent paralysis.

Some colleagues chalked up the entire event to misfortune, to an inevitable component of fieldwork. These kinds of losses are often viewed as the undesirable but unavoidable and acceptable collateral damage that sometimes occurs as we work to understand and promote the well-being of a population, a species, an ecosystem. Nothing could be done for the fawn now. I knew that. But I would forevermore carry her loss with me.

On the rare occasion when I encounter another who has experienced a similar grief, someone who has confronted loss and is not unmoved by the costs despite recognizing the value of the work, I am comforted. I feel hopeful. Because if our discipline is one that records such losses as inconsequential—little silent blips on the loud path to scientific success—then what can be done for us?

Courtesy of Marie Tosa

Marie Tosa found her passion for wildlife during high school, while working with bats in New England. She received her bachelor's degree in biology and environmental studies from New York University, her master's degree in zoology from Southern Illinois University Carbondale, and her PhD in wildlife sciences from Oregon State University. Marie is now a research scientist with the Washington Department of Fish and Wildlife working to preserve species of greatest conservation need.

Originally from eastern Massachusetts, she has slowly made her way westward across the United States while working with awe-inspiring species in amazing places like Yellowstone, Grand Teton, and Glacier National Parks, and the H J Andrews Experimental Forest in Oregon. Marie, her partner, Scott, their daughter, Mae, and their Australian cattle dog, Blueberry, love to go on adventures among the old-growth trees in Oregon's mountains and frolic on its beaches. You can also find Marie tending her fruit and vegetable garden or paddleboarding during the summer.

The Biologist I Wanted to Be

Dianne Van Dien

Just as the dust devil hit, I turned my face and clutched my clipboard. Dirt and tumbleweeds pelted me while the burst of wind tore at my data sheets. The mini tornado passed quickly with a force I hadn't expected.

I was in the midst of a six-month internship as a field technician studying the impact of National Guard training maneuvers on golden eagles and prairie falcons in the Snake River Birds of Prey National Conservation Area. Just south of Boise, Idaho, this area is home to the largest nesting concentration of raptors in the Western Hemisphere. At the time (1994), it boasted forty-four pairs of golden eagles, more than two hundred pairs of prairie falcons, and hundreds of other nesting raptors representing at least a dozen species.

Ever since reading Farley Mowat's *Never Cry Wolf* at age twelve, I'd dreamed of becoming a field biologist. I imagined a future where I, like Mowat, would spend months in remote country, braving the elements and watching wild creatures to discover unknown details of their lives. Fourteen years later, when I found myself in the vast high desert of Idaho listening for the radio transmitter signals of passing prairie falcons, I thought my dream had come true.

THE ROUTINE

During my first two months of work, the permanent biologists took me and two other newbies under their wings to teach us how to use radio-tracking (telemetry) equipment and how to navigate between the river canyon and the surrounding high plains. I observed cliffs to determine where pairs of eagles or falcons were nesting and provided an extra pair of hands when the biologists attached radio transmitters to the raptors we trapped. Later, when the field season was in full swing, eight experienced seasonal technicians joined the crew. Each week, half of us were assigned to monitor nesting activities and half to track the radio-tagged raptors.

Members of the tracking team were deployed across the open sagebrush country at predetermined locations—often on top of buttes because transmitter signals could be detected more easily from high points. We'd leave as a group at 5:30 a.m. towing a trailer loaded with ATVs, and each of us would either be dropped off directly at a tracking site or would ride to one on an ATV. We'd then spend eight hours swiveling our antennas in search of signals while sitting on a lawn chair with a 360-degree view of the middle of nowhere.

As I stared at the vast landscape, I could see no one, but the on-and-off chatter coming from my handheld radio told me my coworkers were out there. Occasionally I'd spot a raptor or a pronghorn or a jack rabbit. But mostly my only companions were sagebrush and saltbush, cheatgrass and Russian thistle extending to the far horizon, along with endless wind and sun. I rarely even saw the falcons that we tracked as they left their canyon nests and hunted over the adjacent open country.

When anyone heard a transmitter signal, they'd announce it over the radio: "I hear 223." And everyone would tune their receivers to that frequency. If three or more of us heard the signal simultaneously, we'd each take a compass bearing on the direction it came from and radio it to the team leader, who'd plug the bearings into a computer program that would triangulate them to pinpoint the falcon's location. It was an ingenious setup to get real-time data on bird locations. If a lot of falcons were out hunting, these tracking days were busy. But when the birds weren't flying, the days were quiet and slow, leaving a lot of time for reflection.

On nest-watching days, we'd each drive into the canyon alone, then observe a nesting pair from a blind and take notes on the birds' behavior. Often one parent sat on the nest while the other perched nearby or hunted in the vast domain monitored by the tracking team. I'd spend the day staring at the nest through a spotting scope and writing down a code every thirty seconds that indicated what the birds were doing. Usually it was a shorthand version of: incubating, incubating, incubating, preening, incubating, incubating, turning the eggs, incubating, incubating . . . hour after hour. But then the falcon on the nest might start vocalizing—a series of short *kek-kek-kek*s. And its mate would suddenly appear with part of an antelope squirrel, drop it off at the nest, and leave. The incubating parent would eat and go back to incubating. Twenty seconds of excitement and the rest of the day: incubating, incubating, incubating . . .

THE TREASURED

Even though these uneventful days could get quite boring, watching wildlife was incredibly rewarding. Over time, I began to see "my birds" as individuals, each with its own behavioral quirks. And every so often I witnessed the spectacular. In those moments, nothing else existed.

Early in the season, I was tasked with looking for a golden eagle whose transmitter revealed that the bird wintered in Idaho's National Guard firing range and summered in British Columbia. With satellite transmitters still in their infancy, finding the eagle was part of a study to ground-truth the accuracy of the satellite information by checking how closely it matched the bird's actual location on the ground.

When I found the eagle, she was about fifty yards from the road, standing on a rock surrounded by sagebrush. Between her feet was part of a rodent or rabbit, and she repeatedly bent forward to tear off and swallow small morsels of it. As I watched, I glimpsed the bushy tail of a coyote meandering through the bushes. The canid walked right up to the eagle, and for a moment the two stared at each other. Then the coyote lunged forward; the eagle jumped, spread her wings, and extended her talons. The coyote cowered—but just slightly. A second later, the eagle flew to a taller rock about a hundred yards away. Grabbing the remainder of the eagle's breakfast, the coyote disappeared. The eagle spent the rest of the day on the rock, preening and staring out over the plains.

Another time, I watched a pair of prairie falcons exchange prey in midair. The male flew into the nest territory, holding a small ground squirrel in his talons and *kekk*ing to announce his arrival. The female *kekk*ed back and left the nest. Pumping her wings, she sped toward her mate. He let the squirrel fall, and a nanosecond later the female snatched it from the air with her feet. She carried it to the nest and fed it to the young.

One evening at a different nest, the male falcon flew in to take his turn incubating, but his mate would not budge. He nudged her repeatedly until she finally stood up and flew off. I later witnessed this male coax his mate off the nest on several other occasions. By monitoring different nests numerous times throughout the season, I learned that each pair interacted in its own way.

Even when my study birds spent the entire day preening and incubating, I might still glimpse a lizard as it dashed over the blind. While sitting in a lawn chair listening for transmitter beeps, I might see a badger amble past or a northern harrier perform its undulating sky dance. Or I might hear coyotes howling at sunset and then spot them on the cliff edge, silhouetted against a magenta sky—mom, dad, and pups, all with noses pointing up to the stars.

THE UNCOMFORTABLE

When it came to the fieldwork, I had no trouble with the fifty-five-hour work weeks, the low pay, or spending so much time alone. I even grew accustomed to squatting to pee out in the open. What I did struggle with, however, was being cold for ten solid hours. In March, the temperature rarely rose above 40 degrees Fahrenheit. Sitting out in the open and being pummeled by constant winds chilled me to the bone. I sipped from a thermos of hot tea throughout the day and ate more snacks than I needed, causing me to gain ten pounds. My muscles ached from shivering.

A brief respite came as the seasons changed, but soon the days were scorching. My hair bleached to blonde, and my skin tanned even under layers of SPF 40 sunblock. My body felt gritty with salt from sweat that evaporated so quickly I didn't even realize I was perspiring. One day I drank two gallons of water and never had to pee.

Worst of all were the driving hazards. Roads with rocks jutting everywhere scared me silly. I constantly feared I'd get a flat tire. One road made a 90-degree turn along the edge of a cliff, and while approaching the turn, all I could see over the hood of the truck was the river flowing fifty feet below. That was scary enough under dry conditions. The day I had to make that turn with three inches of snow on the ground, I nearly threw up.

Operating an ATV filled me with a different sort of anxiety. Although others seemed to relish speeding over rocks and swerving around bushes, I maneuvered my machine with the slow steadiness of a tortoise, all the while filled with guilt and worry that I might be keeping my coworkers waiting.

THE GATE

Occasionally we were assigned to follow golden eagles as they hunted so we could record the location and habitat where they made a kill, as well as the type of prey they'd taken. But I was never very good at this. Not only was I a cautious driver, but also the eagles' flight paths never aligned with the main gravel roads. To follow an eagle, I usually had to navigate a network of unfamiliar and unmarked two-track paths in flat, open country with few landmarks. Too often the road I chose didn't take me where I needed to go. But one day I was actually keeping up with an eagle and feeling quite pleased with myself. *Finally I'll get this*, I thought. And then I came to a gate.

We were in cattle country, so if we opened gates, we were supposed to close them behind us. The gate I encountered was a stretched-wire gate, made from four strands of barbed wire pulled taut across the road. At one end, the barbed wire was fastened to a moveable fencepost. That post was held in place by two wire loops connected to a stationary post. To unlatch the gate, you had to slip one wire loop off the top of the moveable post and then lift the post out of the bottom loop. You'd then drag that post, and the now-sagging barbed wire, out of the way so you could drive through. In theory, opening and closing the gate was simple. In reality, some gates could be stubborn. This one seemed impossible.

After struggling for what felt like eons to slide the wire loop off the top of the post, I finally succeeded. I moved the gate out of the way and ran back to my truck. The beeps coming from my telemetry receiver told me the eagle was still within range. Relieved, I drove through.

But if opening the gate was a struggle, closing it was a battle. I simply wasn't strong enough. I pulled and pulled, trying to position the post so the wire loop would slide over the top, but I couldn't quite do it. As the beeps emanating from my receiver grew fainter, I contemplated leaving the gate open and coming back later. But I didn't want to risk angering a rancher, so

I kept brawling with the gate. I changed my grip, pushed and pulled from different angles, threw my whole body into it. Nothing worked—until finally it did.

Elated, I turned to run to the truck but was jerked backward. My head was attached to the post. Part of my hair was stuck in the wire loop—and not just a few strands. A four-inch section was caught at chin length. Not wanting to lose more time, I grabbed my hair above where it was stuck in the wire and tore it free, leaving behind a chunk of blonde tresses. With my new lopsided hairdo, I raced back to the truck. The eagle's signal was extremely faint. I drove in the direction the bird had been traveling, but I couldn't catch up. The signal faded completely, as did my spirits. I felt like a total failure.

When I look back, neither the gate nor my other struggles seem reason enough to have believed I was no good at fieldwork, but at the time I doubted my abilities wholeheartedly. It didn't matter that other people got their trucks stuck and had to be towed. It didn't matter that my observation skills improved as the season progressed. (In the beginning, even with the help of telemetry, it had taken me several minutes to find an eagle on the cliffs; later I could spot one within seconds.) It didn't matter that I'd navigated back to the office on unmarked, unfamiliar roads in dense fog when no landmarks were visible (while the biologists were expecting a call for help). No one ever told me to drive faster or to do a better job. Nevertheless, I felt inadequate. That gate was the last straw.

I loved Idaho. I loved the birds. But when the field season ended, I went back to working in environmental education, which gave me a sense of purpose and let me serve the wild creatures I loved without suffering in difficult temperatures or driving in scary conditions.

FAST-FORWARD TWO DECADES

Blue Springs, Missouri. I'm sitting at the top of a small dam on the edge of a lake, watching sixty-three turkey vultures. Every day around sunrise, the vultures fly from their nearby forest roost to perch on the rocks along the lakeshore. They remain there for two to three hours, preening and interacting while waiting for the sun to warm the morning air and create thermals for soaring. Every other day I wait with them.

Observing turkey vultures is not my job—but it has become my passion. When I moved to Missouri several years ago and discovered two easy-to-access vulture roosts not far from my home, I began monitoring them—observing the birds' behavior and recording their changing numbers through the seasons. The more I've watched and studied vultures, the more I've realized there's still much to learn about them.

When the air becomes ripe for soaring, the vultures embark on a day of foraging. Later they will go their separate ways, but during takeoff the vultures rise one by one to glide and circle together. I call this moment the fly out. They pass below me, above me, and at eye level. On windy days they sail back and forth, swooping over the dam, sometimes zipping past me, tilting, and playfully chasing one another. On humid days when the air is heavy and slow, they flap a few times and float silently by. Although I've seen vultures do this a hundred times, I never tire of watching them: six feet of winged elegance expertly harnessing the air. They circle gradually upward and disperse out of sight. Then I return home to begin my own day of work.

It took me a long time to stop feeling disappointed in myself for not becoming the biologist I had wanted to be—someone who could go it alone in the wild, no matter how rugged the conditions. I will never be the biologist my younger self envisioned, but when opportunities arise to go into the field for short stints, I take them. My time watching vultures—sitting near their roosts, getting to know their routines and those of the wildlife that surrounds them—lets me witness nature's marvels on my own terms. I record what I see and share what I learn with others. In the process, I've discovered something my younger self didn't know: there's more than one way to be a biologist.

Dianne Van Dien grew up near the Appalachian Trail in the New York and New Jersey area. Daily walks with her grandmother, looking at mushrooms, bugs, and salamanders, set the stage for her lifelong interest in wildlife. Her career began with a series of short-term positions, including seasonal

Don Shelden

naturalist at Custer State Park, educator for Hawks Aloft, Inc., and aviculture intern for The Peregrine Fund in Hawaii. After nearly a decade as a lab technician at the San Diego Zoo, she returned to school and earned a master's degree focused on environmental communication and writing.

Today, Dianne works as an editor for the Missouri Department of Conservation and writes the Nature Lab section of the *Missouri Conservationist*. She enjoys participating in annual bird and butterfly surveys and pursuing her personal goal of learning as much as she can about the vultures in her corner of the world.

A Natural History of Gestation and Birth

Marlene Farrell

Snorkeling surveys taught me how unlike a fish I am, especially one fall when I was four months pregnant with my second child. I was counting fish for the US Forest Service in the foothills and mountains of north central Washington. In everything from small creeks to boulder-strewn rivers, we were documenting fish diversity in relation to natural and human-caused disturbances.

To prepare to submerge myself in frigid rivers, I bundled first in head-to-toe fleece. These layers fit underneath a heavy-duty canvas dry suit with rubber seals that bound my wrists and neck. My belly was not too large yet, but it was beginning to press against the unyielding canvas. Another crew member had to open and close the zipper in the back, and once inside the dry suit, I felt less human than vehicular, especially when I added my accoutrements: snorkel, mask, PVC arm cuff with attached pencil for recording fish sightings, and waterproof headlamp. The headlamp both lit my way and illuminated the fish during nighttime surveys. In the darkness, the fish flashed and darted rather than lurking out of sight under and within submerged woody debris.

We entered the water as if receiving a group baptism. But rather than leaning back to immerse ourselves, the four of us spread out across the river, dropped to our knees, kicked the toes of our boots into the gravel, and plunged face first into what was often a powerful onslaught of cold water.

We then worked our way slowly upstream, observing fish within our parallel lanes.

Under the surface, the water was hazy, almost viscous, and the colors muted. I lumbered against the flow, my gloved fingers making small purchase in the streambed, sometimes sending pebbles tumbling past me. With no smells to distract me and the sound of rushing water an internalized white noise, it was easy to focus on the fish—dark shadows that suddenly glinted in my light. Examining this unfolding world, I barely noticed my face slowly numbing in the icy water.

I had to search the substrate for some of the fish, like the sculpins, which rested on the streambed and blended in, matching the mottled patchwork of tans and grays. To find them, I learned to spot the frilly edges of their fan-like fins. The trout and whitefish, stationed mid–water column, let us creep toward them. Surely, they heard us coming. But they held steady against the flowing water, hovering in place with little apparent effort. They scattered only when I came close enough to touch them.

These beautiful fish, with sleek bodies that have evolved over millennia, moved with a finesse I lacked. I needed three points of contact—both boot tips digging into the streambed and a gloved hand gripping the substrate—to hold my position. The little one inside me, however, was something in between. She was suspended in water, moving tiny limbs and spinning in place. Unable to undulate or dart, she was more like a goldfish trapped in a bowl. But she was in her natural habitat.

We finished snorkeling a section of river and emerged for a break on the bank. Now that I was upright, the small bulk of her pressed on my bladder. I asked for help unzipping my dry suit, then squeezed my head through the neck gasket and peeled the unwieldy suit most of the way off, only to reverse the arduous process a few minutes later. I envied my colleagues who went about work without needing double pee breaks or cycling through hunger, eat, heartburn, repeat.

When we returned to the water and I thrust my face below the surface again, the force of the current hit me anew. The full body pull exerted by the river reminded me suddenly of twenty months earlier, when my first baby was tugged from my body, and I, under lower-half anesthesia, felt my body resist the doctor's efforts to wrestle her from my womb. How I got there was closely intertwined with my previous field job, which was as terrestrial as this one was aquatic.

A couple of years earlier, I had worked on a barred owl project in Washington's Cascade Mountains until mere hours before my baby was born. Having trapped and radio-tagged several owls, my coworkers and I were tracking their movements and determining whether they had successfully raised young.

The barred owls hadn't been easy to trap. The wily birds had moved into the region after people had altered the landscape, and now they were slowly outcompeting the native spotted owls, which were listed as threatened under the Endangered Species Act. Despite the barred owl's status as a "problem" bird, I admired how these feisty balls of fluffy feathers saw through our plots to trap them. They ignored the recorded calls we played to draw them into our capture nets, and they failed to become entangled in the nets we placed around Bob—a bobble-head great horned owl—to provoke their territorial dive-bombing attacks. It wasn't until we tethered skittish mice to a stump near an almost-invisible fine-mesh mist net that we finally caught barred owls. After quickly fitting them with radio transmitters that they wore like little backpacks, we set them free.

Although pregnancy didn't stop me from doing fieldwork, it was a nuisance at times. I blamed my bouts of nausea on the double whammy of long truck rides on bumpy roads and the stench of bug spray blended with the overly strong scent of my coworker's coffee. Flinging myself out of the truck as soon as we reached our destination, I would savor solid ground and gulp fresh air.

Hiking uphill through brush would leave me feeling leaden. I wasn't very forgiving of myself during that first pregnancy. Being a marathon runner, I gritted my teeth and tried to keep up. I was embarrassed by what seemed to me like unreasonable fatigue caused by a peanut-sized being inside me. My male coworkers knew I was pregnant, but I didn't want to use it as an excuse to do less than my share.

My boss came out in the field with us at times. Late one night, we were taking down the mist net after an unsuccessful trapping attempt when we heard a buzzing sound. We had roused some sleepy wasps that nested in a nearby log, and they headed for the nearest target—me. Three wasps stung me as we fled back toward the Jeep. One flew up my pant leg, so without thinking, I dropped my pants down to my ankles as my coworker shooed it away.

I'm not usually allergic to wasp stings, but the chemistry of pregnancy is full of surprises. Suddenly, hives started blooming all over my skin. The pain of the stings gave way to incredible itchiness. My boss, usually the poster child of safe driving, drove us back to town at breakneck speed. My coworker, who had an emergency medicine background, worried me more as he leaned over me wielding an EpiPen and nonchalantly asked me how my throat was feeling.

I was fine after a visit to the local emergency room, a dose of antihistamine, and some observation time. It was, however, a forewarning that pregnancy would take away a bit of my usual rough-and-tumble hardiness when it came to fieldwork. I needed to be more vigilant, more careful.

Sometimes the trouble wasn't physical, it was emotional. My job required me to think and act rationally, but little things like making a wrong turn on a dirt road or hitting my head on a branch in the woods could send me veering down an unfamiliar path of weepy frustration. In part, I was dealing with a bombardment of fluctuating hormones. However, in retrospect, I also think that while we soon-to-be mothers are hosting and giving new life, we are also losing a bit of ourselves in the process. That's scary—and it doesn't end with pregnancy.

On one of my last scheduled workdays, in early winter about twenty days before my daughter's due date, we were tracking the owls through the forest as usual. My new field partner, Maryellen, was also an expectant mother and becoming a friend. We had skied into our project area and located one of our owls. It was a short ski back to the truck. The snow had an icy sheen to it, having warmed and refrozen over old snowmobile tracks. Skiing in had been fine, and as a seasoned Nordic ski racer, I didn't worry about the slight downhill on the way out. At one spot, I swerved left, making a bit of a snowplow with my skis to turn and avoid some icy bumps. Ahead, in the shelter of an overhanging tree, the ground was bare. I reacted quickly by lifting my left ski to avoid the dirt, but I was a moment too late. The ski came to an immediate halt, while I did not. There I was, dazed, laid out flat on my stomach, my face scraped from hitting ice.

And then I felt a gush of warm water that soaked my pants. I jumped up, struggled out of my skis, and yelled to Maryellen, who was just ahead. I ran, feeling a rising panic. She hadn't seen me fall, but she took one look at my face and knew something was wrong. Tossing our gear into the backseat, we were off, driving faster than regulation allowed for the second time as we sped down the winding two-lane highway.

I called my husband, Kevin, on our rarely used, work-issued cell phone as we drove, trying to convey my situation through spotty reception. Twenty minutes later, I was handed off to Kevin, who then whisked me to our nearest hospital, half an hour away. Despite the fear in my gut, we first had to fill out paperwork. After an examination, the doctor decided that an emergency C-section delivery was in order. Until that day, I'd experienced a near-perfect pregnancy, enjoying how my body was fulfilling one of its natural roles, just like the organisms I'd studied. But everything can change in an instant.

A C-section, for me, meant being on the wrong side of a curtain. I knew what was happening on the other side, but I was shielded from observing, which felt like a punishment. Some violent but pain-deadened tugging yielded my baby—purple transforming to pink, taking first breaths, set warm on my skin. Just for a moment. Then I had to be stitched up and wheeled to recovery. My fear and shock and distress at being separated from my new baby soon yielded to breathless wonder.

Back in the cold, dark river in early fall, I was again deeply enthralled, as I worked with single-minded focus to document the composition of a freshwater ecosystem. During breaks, I could contemplate baby number two and how my life was about to change once more. But I still had five months to enjoy her as my internal guppy. I could feel her animated stirrings, her tumbles and pokes, while I scrabbled in rivers and, later, did office work. When her movements quieted for a little too long, we rushed to get an extra ultrasound. I need not have worried. Soon thereafter, she was born in perfect health, and I had the joy of fully participating.

My new role as a mother of two seemed like a natural extension of my work as a biologist: observing life cycles and behavioral and physical adaptations, and weighing the shifting balance of ecological costs and benefits. Only now, my perspective came from inside the human life cycle. My babies were my very own study species—and later became my field assistants. I saw everything with fresh eyes—through their eyes—as together we marveled at the elegance of water sliding over rocks, the intricacy of a feather dropped in the woods, the splendor of a world that began right outside our back door.

Tough Girls

Isabeau Dasho

I know that many women who do fieldwork have to fight for their place in the truck, or prove themselves by doing endless grunt work, or deal with a low hum of sexual harassment from coworkers. The field can be ripe for abuse and isolation, the message of conservation—of living in harmony with our ecosystems—always and endlessly in conflict with the macho myth of the wild, where Teddy Roosevelts perpetually pit their brawn against nature. I know many women who have stories of proving themselves again and again, of staring down fierce predators as well as indifferent field teams. This, however, is not one of those stories.

As a high school intern for the Missouri Department of Conservation under the tutelage of a famous crayfish biologist, I was insulated from most of those challenges and obstacles. My mentor was a happy boss with many talents, not the least of which was his ability to build great teams. He was aware of and concerned about the plight of women in research and conservation generally. His allyship included both action and reflection, evident in the teams he built, which always included women, and in how those teams functioned. During my internship doing stream habitat surveys and crayfish population counts, my full participation in the work was always encouraged, mentored, and praised.

Our team leader that summer was Adam, a kind and affable man who wore responsibility like an easy coat. He took his cue from our boss, and his style was simple. Get the work done and be chill. We also had Ed—an unusually silent yet dependable crewmate who spoke only a handful of words to me all summer even though we shared a near-death experience—and Janna.

This is a story about Janna.

The first thing you should know about her is that she had a smile as wide as the Nebraska sky. Big blue eyes and long blonde hair to match. She looked like she could have been a prom queen in a John Hughes movie, or a coquettish farmer's daughter in an old musical. She was lovely, approachable, and tough as nails. The kind of person one is both surprised and not surprised to meet in fieldwork. A sort of Rachel Carson Disney princess, she looked like she belonged in an enchanted wood.

Janna's outward loveliness was surpassed only by her ebullient personality and unassuming good humor. An essential asset to the team, she was hard to ruffle and didn't ruffle others. Crack a joke and she'd laugh, show strain and she'd offer to help, lose heart and she would carefully and kindly encourage you.

As an awkward seventeen-year-old intern, I looked up to Janna, who came to represent a kind of unflappable and knowledgeable crayfish heroine to me. I admired her, obviously, but with the hindsight of years, I see that I also had a bit of a hero-worship, platonic crush on her. I didn't want to be with Janna romantically, but I did want to *be* her. Watching her haul diabolically unwieldy and heavy quadrats for capturing crayfish up and over small waterfalls was an invitation to be stronger than I thought I was. If we found a venomous cottonmouth snake in one of the crayfish traps, it didn't faze her. Oftentimes in that fraught moment between surprise, terror, and curiosity, you could almost see a dare form on her face. I don't know who she was daring, but I wanted to show her I was up for it. I wanted to be the kind of girl who was up for it.

Missouri in the summer is a particular kind of brutal. The temperature is a sweltering midnineties by ten in the morning, and the air is so thick with humidity that it feels like you're walking through diaphanous hot butter. Your skin is instantly slick with sweat as you plod from air-conditioned building to hot-ass car and then wait for the air conditioning to kick in. It is a distinctive torture. And it's the only time the Missouri Department of Conservation wants to survey for endemic and endangered crayfish in the state's storied Bootheel, where the critters live.

We were midway through our field season, and it was misery.

Rural Missouri offers few places to stay, but we did find a godforsaken murder hotel off the main drag of a town whose name I can't remember. And as you can imagine, there wasn't much for entertainment other than planning our work for the next day. We combed through maps of mostly unmarked dirt roads to locate the best access point for this one branch of this one tributary in this one watershed where we would survey the endangered crayfish species that called this slip of water home. The next morning, we had to be on the road by seven at the latest because the mid-July heat was rocketing upward. Field work left much to be desired for me that summer.

As miserable as the heat is—and it is, I cannot stress that enough—southern Missouri is beautiful in the summer. It's a riot of greens—the dark verdant undersides of leaves in the gullies and ditches, the living breathing emerald of willows that overhang the creek beds, the light olive-green of dry grasses on the roadside. Overbright sunlight dances in the thick air, shafting through the canopy and catching wisps of wings and particles between the shadows. It's a mystical place, removed from time and technology. At the best of times, being in southern Missouri felt like walking into a cathedral—the creek the center aisle and the downed trees the pews. We were careful parishioners, prayerful in our way, doing good and needful work. And in those moments, even the heat seemed to subside, offering a dazzling second of pure rapture.

When we were told we had to haul the crayfish traps up over the boulders above the waterfall, the peace splintered and the work began. I remember that morning clearly because it was so beautiful, and it went so beautifully wrong.

Janna was ahead of me with equipment in her arms, her long ponytail swinging jauntily behind her. She had a happy walk even when carrying a massive amount of heavy and awkward gear. I, by contrast, was puffing and wheezing with my own traps, stumbling over rocks and trying to pull my own weight against gravity. We carried on this way for a few yards, Janna a sturdy and indefatigable crayfish fairy and me her haphazard sidekick, when I heard her softly curse, "Motherfucker." She'd paused in her ascent of the boulder face and was looking down at the back of her knee.

I could already see a nasty welt forming. I'm sure I said something brilliant, like "Yikes!" or "That looks like it hurt!" Janna smiled and turned back to her task, shaking off the sting like a horse shakes off flies. I was, of course, deeply impressed. I filed this toughness away and vowed to stop whining about the heat, even to myself.

We got to the top of the falls, set the traps, did some swimming to cool off, and then pulled out our lunches. By now about two hours had passed since the sting, and Janna's welt had doubled in size. This was novel to all of us, three twenty-somethings and a teen. I believe we made a wisecrack or two about the welt's size as a way of voicing our concern. Because Janna minimized her own hurt, we felt comfortable doing the same. She used some ointment from the first-aid kit on the swelling, and that was that.

By afternoon, her leg was clearly nothing to joke about. While Janna carried her share of the gear back to the trucks, declining to accept help from us, her gait was already stiff. In the truck, it was clear that her leg was making her uncomfortable. She silently repositioned it with her hands as it continued to swell. By the time we got back to the hotel, we were alarmed.

It's worth remembering that we were all young idiots. I'm sure it's clear to any casual reader that Janna was having an allergic reaction to the sting and that at this point, given the visible swelling of her limb, her pain, and the marked decrease in her mobility, we should have taken her to a clinic.

But the conversation we had as a team went like this: we were in rural Missouri, the nearest clinic was hours away, and it would be closed by the time we got there. The nearest hospital—and Janna would not even entertain the thought—was likewise hours away. We had only one workday left before heading back home to civilization and the clearly needed medication. The survey we were working on was crucial. This crayfish species is found in only two tributaries, and information about its populations is vital to its protection. You can't protect a species if you know next to nothing about it. If Janna thought she could tough it out, who were we to question her?

I hear this conversation so clearly in my mind, and I have thought of it often in the intervening years. Adam, as team leader, was in a tough spot. He'd known Janna for a few years, knew her capabilities, knew her strengths, and, I imagine, knew her stubbornness. In many ways, their familiarity with and trust in one another was deep, and their trust functioned without words. Often, they could sense any frustration in each other and easily diffuse it with banter or a wide and knowing smile. Adam trusted Janna to know her limits. Or maybe he trusted himself to get her through the last day of work. As both a project leader and a friend, Adam was in a bind.

And now I think of Janna, too, in a way I didn't really have the capacity for at the time. Janna was used to being the only woman on the crew. And while Adam and Ed were easygoing, I'm sure the pressure Janna put on herself to show that she'd earned her place was not insubstantial. There's a feeling of responsibility that comes with being an "only"—a pride maybe, a joy perhaps, in being special, but also a weight. There's the burden to do it well, to make sure that whoever comes after you has an easier path and won't have to work so hard to earn her place. I imagine that being a female mentor to me also played a role in what Janna felt she must live up to.

I can see Adam and Janna in the waning twilight of that evening: two coworkers, two friends, each trying to suss out the other. There's a quiet trust in letting someone you care about fail or hurt themselves. Janna, rightly or wrongly, was asking her teammate to trust her, and rightly or wrongly, he did.

We did not head home or to the hospital that night. Instead, we planned to check Janna's leg in the morning, then finish the survey, grab our traps and equipment from the stream, and head home. At Adam's insistence, Janna dutifully drew a Sharpie border around the welt that now encompassed the entirety of her knee, so we could discern any change in its size. This done, we all went to bed.

By morning, Janna had a fully fledged peg leg. It was too stiff and swollen to bend, giving her the ungainly walk of a pirate, with a pink tree trunk attached to her small frame. The Sharpie border from the night before was so twisted and bloated that it had become meaningless. Janna limped out to the truck and opened the door. Staring at the cab, she confronted the conundrum of how to get in when she couldn't bend her knee. Her leg was hideous. The fact that she couldn't bend it was frightening. It was also kind of funny. But we had work to do.

By this point, we all knew we were dealing with a medical situation that required professional help. Adam wanted to leave Janna in the truck, forget the crayfish, simply collect the gear in a speedy jiff, and race to the nearest clinic. When we reached the field site, Janna made it clear she had other plans. The duty she—as one-quarter of the team—felt to us was obvious. She hated to let anyone down, and that feeling of pulling your whole weight for the team compelled her to leave the truck. Despite the obvious encroaching medical emergency, despite our growing worry, despite the knowledge that we could do it without her, Janna hobbled up the trail.

I remember that she was wearing a baby-blue T-shirt and navy-blue shorts that made for a memorable contrast with her red leg as I walked behind her. I remember how Ed offered, in his quiet way, to take some of the equipment she was carrying, and how she politely declined. I remember how Adam, in front, constantly cast his eyes back at us, worried, chagrined, resigned. But mostly I remember Janna.

No longer the crayfish Disney princess, she'd now morphed into Furiosa from *Mad Max*—taking zero shit from any of us. I had never seen so active an example of "grin and bear it." But grin she did, relentlessly, through the heat and along the trail, while carrying the excess weight of her expanding leg and the gear and whatever expectations she had of herself. She pulled hard on the straps of her backpack and hiked onward.

Janna was not the first capable woman I had met. But she was the first capable woman I had seen in action under extreme duress. And on her bum leg, she hiked to the rhythm of that word like her life depended on it. *Cape-ah-bull. Cape-ah-bull.*

The ride back to town was weird, and time seemed to dilate, as happens when you are occupied with worry. We couldn't go fast enough; the red lights were interminable. And once we hit the highway in our government-owned vehicles, we drove the speed limit like the good public servants we were. Janna's leg was splayed out on the dash, fully extended, painful, and hideous to behold.

I'd like to tell you we went straight to the hospital, but once again, Janna had other plans. She had excellent seats at a sold-out show that night with her fiancé. The tickets had been expensive. They loved this band. She would go. And then she'd go to the hospital.

After treatment, Janna's leg returned to its normal size. She finished the field season like a total badass. Our boss was horrified when we told him Janna's story, clearly a bit shocked that dependable Adam had let things progress as they did. Adam, too, was a bit shocked, I think, by the retrospective narrative that we shared in the little office. It was a harrowing tale. Janna shrugged.

I think of that, the way her reaction to minimize her injury functioned like a reflex. Like a smile in a bar when you're uncomfortable, like a little laugh to diffuse tension, or the way you swallow a retort you're poised to give. At the time, I admired Janna's grit, her ability to get through physically, but now I wonder what happened to make her behave that way even among trusted colleagues and friends. Why she wouldn't admit to the danger we all clearly saw or accept the help we all offered. Why at some level she felt like she couldn't, or shouldn't.

And I wonder how we begin to dismantle that reflex for the benefit of all the tough girls out there. For all the girls out there.

Maren Robinson

Isabeau Dasho grew up immersed in the natural beauty of the Midwest, exploring Wisconsin's rivers and hunting crayfish in Missouri during her school years. Two summers as a field intern with the Missouri Department of Conservation taught her that fieldwork wasn't her path, but her love for the science, the work, and the people behind it endured. Moving to Chicago, she witnessed urban environmental renewal firsthand, spotting a beaver family in the Chicago River and hearing coyotes rallying close to her apartment in nearby cemeteries. Inspired by nature's resilience and the shared challenges people face, Isabeau joined the Harris School of Public Policy Writing Workshop, where she helps young policy-makers refine their writing to leverage science and data for the future.

Isabeau's writing has been published in the *Chicago Policy Review* and the *Missouri Department of Conservation Magazine*. She runs a feminist podcast celebrating romance novels.

Close Calls

Just Another Walk in the Park

Libby Ehlers

For biologists working in bear country, bear safety training is as essential as carrying water and food. During mandatory training sessions, I'd distract myself by imagining the cool stories I'd be able to share someday about my close interactions with bears. It was hard to believe I might actually need that training to survive.

I was spending my first summer as a biological field technician working for the Yellowstone Wolf Project. My field partner, Matt, who led the summer predation project, and I were following a radio-collared female wolf—number 593F from the Leopold pack—from May through August to see where she traveled, hunted, and fed. The work was awesome yet tough. We hiked more than eleven hundred miles through the pack's territory that summer, endured intense heat and insects, and bushwhacked through thick stands of lodgepole pine trees that were growing back after the massive 1988 Yellowstone fires. But we were continuously surrounded by breathtaking landscapes, geothermal features, and spectacular wildlife. We spent much of our time in the wildest of places, ones that few people had experienced or explored as extensively as we did that summer.

During the first few weeks of the study, we ran into bears several times. Each encounter taught me more about how to react to these potentially dangerous animals. I met my first grizzly in the backcountry while hiking to one of our study sites. It was standing up on its hind legs anticipating our arrival as we rounded a bend on a game trail in the forest. Clearly, the bear had heard us approaching and was trying to determine what was creating the noise. We halted abruptly upon seeing the young adult grizzly, raised our arms, and spoke in a firm, calm manner, just as we'd been trained.

"Hey, bear. It's okay, bear. It's okay, bear."

"We don't mean any trouble, bear," we continued as we walked backward.

Both Matt and I pulled our bear spray canisters out of the holsters attached to our belts, pulled off the safety clips, and pointed the canisters at the swaying bear. Once we'd put some distance between ourselves and the grizzly, we hiked on, skirting widely around it. The bear was curious but not stressed—a perfect encounter.

On another occasion, we met an adult grizzly as we reentered the forest after walking through a large meadow. We heard tall shrubs rustling about sixty feet ahead of us. Seconds later, a bear exploded briefly into view and then sprinted away, quickly disappearing.

These initial grizzly encounters provided me with critical experiences that reinforced the importance of responding calmly and effectively to bears in the backcountry. Each encounter with a bear is different. Reading animal behavior and then deciding how to react are learned, but essential, skills when working in bear country.

We were tasked with finding wolf kill sites—places where wolves had killed prey—by examining locations collected from a GPS collar worn by Wolf 593F. GPS locations that formed a cluster of points when mapped indicated that 593F had stayed in one area for an extended period, possibly feeding on prey she had scavenged or killed. We checked each of these focal areas to determine what type of carcass, if any, 593F had eaten.

In winter, wolves hunt in packs and prey on large mammals like deer, elk, and moose. During the summer breeding season, though, wolves regularly target smaller prey they can kill while traveling solo or in small groups. Although we often located deer and elk carcasses at focal areas, we'd also search for any other biological signs we could find—piles of fur from the wolf or her prey, beds where wolves had rested, bone fragments from prey—not only to confirm that 593F *had* spent time in a particular area but also to determine *why*. Was she sleeping, capturing and eating a meal, or simply milling around the area? This information would contribute to a long-term study to better understand what wolves eat during the summer when they are raising and feeding their offspring.

Five weeks and hundreds of miles into the study, we were settling into a routine and feeling comfortable with our adventurous job. In preparation for a big day in the field—eighteen-plus miles of off-trail hiking and thousands of feet of elevation change on a scorching hot day—I'd met Matt at four thirty in the morning, and we'd begun hiking soon afterward. By noon, we had covered nine miles and checked most of our focal areas. We had only two areas left to search. Although we had each started out carrying about a gallon of water on our backs, we had been low on water for hours. Stopping at Lava Creek, we took a much-needed break in the shade, filtered water from the creek to refill our water bottles, snacked, and splashed cool water on our faces and hats.

Feeling more energized, we started our climb up the east side of the Lava Creek drainage. Half a mile from our next focal area, we started clapping and yelling "Hey, bear!" to warn any lingering scavengers of our presence. This "be loud and known" protocol was routine for us as we approached our study sites and was especially important when we were engulfed in dense forest. To get to this particular site, we had to bushwhack through a perimeter forest of tightly packed twelve-foot-tall lodgepole pine saplings. As we approached the focal area, we entered a large, roundish clearing. Walking into the beautiful meadow with its lush carpet of tall grasses felt like a breath of fresh air after spending the previous thirty minutes getting poked by pine needles and clambering over fallen trees.

We took a quick look around the grassy clearing where our wolf likely had fed. Matt and I then moved apart, divvying up GPS points from 593F's collar so we could search every spot in hopes of finding animal signs—and of winning our friendly competition to be the first to locate a carcass. (We each carried a handheld GPS unit that had the locations we needed to examine.) Up ahead, I saw a big grassy mound with what looked like sticks protruding from it. Antlers, perhaps? I walked closer to get a better look. As I approached, I smelled the familiar scent of death and saw a huge log lying across the tall grass on the right side of the mound. Just in front of and under the log, I spotted a clump of brown hair that looked like an animal hide.

Strange, I thought. *Why is the hide still here?* Usually, hungry scavengers dragged away the dead animal's hide, and it had been five days since 593F had been at this site.

And then I saw a head. It turned toward me—and charged.

For a moment, time slowed as I tried to process a large bear barreling toward me. I stood my ground and started yelling profanities. The bear reached me in four strides. Screaming, swearing, I frantically waved my arms with my bear spray in hand. I hadn't had time to remove the safety clip from the spray canister but remained strong in my stance, trying to look as big as possible. With only inches between us, the bear rocked onto its hind legs, then lunged toward me. Now face-to-face, I turned away and dropped to the ground in submission, tucking into a fetal position. I felt the weight of the bear on top of me and pressure on my lower backside, and thought, *How bad is this going to be?*

I knew I hadn't yet been ripped open, but I braced for the inevitable, staying as motionless as possible to avoid provoking the bear's instinctual aggression. And then, suddenly, I heard Matt's voice.

"Are you okay? I know you're not okay, but are you okay?"

Matt's words washed over me, telling me I was all right and the bear was gone. Materializing beside me, Matt helped me stand as I shakily found my feet.

"Yeah, I'm okay. I'm okay," I repeated, as if trying to convince myself.

I reached down to my hiking pants. They were soaking wet. "Check my butt . . . am I bleeding? Am I bleeding?" I asked Matt, adrenaline filling my voice with a tinge of hysteria as I looked around for the bear.

"No, I don't see any blood."

"Okay. I'm okay. I'm fine," I repeated. "Where did the bear go?"

Matt reassured me that the bear had left the meadow and headed north. We stood in silence for a few seconds as I took deep breaths, looked around, and contemplated what had just happened. As Matt had run toward me, he'd seen that the bear was a large cinnamon-colored black bear weighing around three hundred pounds.

"Should we get the data from this kill?" Matt's question pulled me back to the present.

"Sure," I replied, "but I'm staying here so I can keep an eye out for the bear." I wanted to remain in the open, where I'd been attacked, to make sure there were no additional surprises.

As we looked around the site, we realized the bear had been eating and guarding a large bull elk that was buried right in front of us under a giant pile of uprooted grasses and other dried vegetation. This cache was about thirty feet across and six feet tall, with two antlers protruding from the pile. Matt climbed onto the top of the cache and pulled on the elk's antlers with all his strength. The mound barely budged, but Matt could see that the elk's neck was still full of meat. This was unusual; we'd inspected dozens of carcasses and typically found only bones and bits of remaining tissue five days after an animal was killed. Because of the crazy heat wave we'd been experiencing, though, the bear might have been feeding only when it was cool, which would explain why so much of the carcass remained intact and still buried to prevent it from spoiling. Given the still "fresh" state of the carcass, we realized immediately that it was dangerous for us to stick around collecting data since the elk would continue to attract hungry and potentially dangerous animals.

"We're out of here," said Matt.

"Great," I said, relieved.

Opting not to call the park's dispatch to tell them about the bear attack since I felt fine and thought I could handle walking the seven-plus miles back to the truck, we took off at a brisk pace and then walked in comfortable silence for about two miles to our final focal area. Oddly, thanks to the adrenaline that still coursed through my body, I felt like we were flying across the uneven terrain.

Once we finished our final site visit, where we found little evidence of our wolf's presence, we stopped to relieve ourselves. I was anxious to finally look at my throbbing behind. As I dropped my pants, I saw a lot of bruising and one spot where the bear's canine tooth had dimpled my skin. After reconvening, Matt commented that being dehydrated made his pee burn. In response to my questioning glance, he said, "I feel like I can tell you anything now that I've watched you almost die." I chuckled, appreciating his subtle effort to lighten the mood.

As we made our way north along the Lava Creek drainage, Matt and I chatted and joked, distracting each other from our recent trauma. We also talked about the attack and analyzed what, if anything, we could have done differently. We felt confident we had done our best to approach the carcass site safely by making our presence known with a lot of noise. The sleeping bear just hadn't heard us. Maybe if we had been together, instead of splitting up to investigate our wolf locations, the bear would have been intimidated and moved off right away. But maybe not.

We were physically and mentally drained when we arrived at the truck. Matt offered to call our boss, Doug, to fill him in on what had happened so I could pick up my dog, go home, and have a beer. I'd planned to go into the field with Matt as usual the next day. But after dinner that night, my phone rang.

"You are *not* going into the field tomorrow." It was Doug. His tone was firm and serious. "You are taking the day off. Take the weekend off. You just got attacked by a bear, Libby!"

"You need to go to the clinic to get checked out and get on some antibiotics," Doug continued. "Bears' mouths are filthy—dirtier than cats. Especially when they've been feeding on a rotting carcass."

"Hmm, good point." That hadn't occurred to me yet.

Instead of meeting Matt the next morning, I slept in and then slowly made my way to the Mammoth Clinic. I was met with astounding enthusiasm because of my injury, and I could tell that my story would spread fast, even if I didn't want it to. The doctor identified some linear bruising along my ribs, likely where the bear had swatted me with its paw, and two puncture wounds. The punctures aligned with where one upper and one lower canine tooth had sunk into my gluteus maximus muscle. The bear had had my entire right bum cheek between its jaws! Realizing this, I suddenly viewed the bear that had attacked me as a really good bear, since it had merely mouthed me, giving me a warning bite rather than a more serious, flesh-tearing bite.

I recuperated over the long weekend, snuggled with my pets, and talked about the attack only with my partner, Nick, who lived in California. Come Monday, I was ready to get back to work, and headed into the field with Doug. I led, maneuvering us first through a dense stand of pine saplings and then through an area laced with hidden and dangerous geothermal features. As we approached my favorite old-growth lodgepole pine stand, I was relieved to be out in the open again. We walked freely through the open understory, awed by the impressive 120-year-old giants. I felt mentally and physically strong as I navigated us through the complex and striking landscape. It felt great to be in the backcountry. Until we reached our first kill site.

"Holy shit! There's a *ton* of bear sign here!" Doug said excitedly, possibly forgetting my recent traumatic event. "There's scat everywhere, and beds, too. . . . Bears were just here."

I froze in my tracks. It was like I had hit a wall. Stunned, my mind and body felt paralyzed, and my confidence plummeted.

"Doug," I finally said. "I can't do this right now. Can you collect the data we need from this carcass? I'll stand up on this log and watch for bears."

Over the next couple of weeks, I took field days in stride once again and started to regain my nerve around carcass sites. However, I struggled with flashbacks. Two weeks after the attack, Matt told me that he and Doug wanted to revisit the attack site (to better understand what had happened there and to finish our data collection) and asked if I wanted to come with them. I felt a surprisingly strong urge to join them on the adventure. After hours of hiking on yet another blisteringly hot day, we arrived at the site, drenched in sweat. As we approached the scene of the attack, adrenaline rushed through me. Everything looked so familiar—like we'd never left.

The three of us approached the enormous pile of dried vegetation that remained from the now-excavated bear cache. A variety of animals had fed upon the massive, eight-hundred-pound bull elk, dragging the limp body as they fed. The carcass had ended up thirty feet behind the log where the bear had been sleeping when I surprised it.

Excited to see where the bear had been slumbering before my arrival, I took a closer look. The bear's daybed was impressive—a cozy two-foot-deep-by-five-foot-wide hole dug out under the shade of the large fallen tree. I pretended to be the bear, crawled into the bed, and took some photos. Afterward, I walked over to the exact spot where the bear and I had gone down in tandem. The spot looked both familiar and vibrant. Sun slanted through the trees, and logs were scattered in an artistic crosshatch pattern over the green grasses that carpeted the ground.

TOP: *A pair of wild harpy eagles in Panama* (Photo by Angel Muela) **BOTTOM:** *Lisa Koitzsch celebrates the perilous rescue of a bison calf.* (Photo by Ky B. Koitzsch)

TOP: *A jackal approaches white-backed and Rüppell's vultures at a carcass in Ol Pejeta Conservancy, Kenya.*(Photo by Darcy Ogada) **BOTTOM, LEFT:** *Darcy Ogada holds a hooded vulture in Jaldesa Community Conservancy, Kenya.* (Photo by Martin Odino) **BOTTOM RIGHT:** *Veterinarians examine a yellow-naped amazon parrot nestling in southeastern Guatemala.* (Photo by Sophie Osborn)

TOP: *A golden-hour glow settles on a remote Alaskan field camp a week before a devastating spring flood.* (Photo by Lindsay Gray Carlson) **BOTTOM:** *Lindsay Carlson estimates the likely hatch date of a brant (goose) egg, using a toilet paper tube covered in duct tape to candle it (use light to check whether it is fertile) on Prince Charles Island in Nunavut, Canada.* (Photo by Paul A. Smith)

TOP: *Erin Stahler and Elise Loggers make the arduous snow crawl toward wolf 1117M.* (Photo by Kira Cassidy) **BOTTOM:** *Erin Stahler with immobilized wolf 1117M of the Cougar Creek pack in Yellowstone National Park* (Photo by Kira Cassidy)

TOP: *Émilie Desjardins pauses between field sites to contemplate the vast Arctic landscape.* (Photo by Marie-Pier Poulin) **BOTTOM:** *Gilles, a bull muskox and data savior, in the polar desert of Alert on Ellesmere Island, Nunavut, Canada.* (Photo by Émilie Desjardins)

TOP: *Susie Adams snorkels for bull trout in Dewey Creek in the Payette National Forest, Idaho* (Courtesy of Susan Adams) **BOTTOM:** *Southern Appalachian brook trout display spawning colors in a mountain stream in Great Smoky Mountains National Park, Tennessee.* (Photo by Bryce Gibson)

TOP: *Zanethia Barnett prepares for a hookah dive to collect crayfish in the Tennessee River, Alabama, while assistant Mickey Bland looks on.* (Photo by Jeffrey T. Garner) **BOTTOM LEFT:** *Zanethia Barnett in her peaceful place—scuba diving with bass and sunfish in Lake Keowee, South Carolina* (Photo by Stacey Wilkinson) **BOTTOM RIGHT:** *Gina Piazza and David, a local student, settle in for a 24-hour stay in a wildlife blind near Okakarara, Namibia.* (Courtesy of Gina Piazza)

TOP LEFT: *Male Guianan cock-of-the-rock in Sentier des Coqs-de-Roche, French Guiana* (Photo by Cody Limber) **TOP RIGHT:** *New Mexico meadow jumping mouse in Apache-Sitgreaves National Forest, Arizona* (Photo by José G. Martinez-Fonseca) **BOTTOM:** *Dede Olson and Mike McDowell in 1980 at the Voltzberg—aka The Rock—which towers above the rainforest canopy of the Central Suriname Nature Reserve* (Courtesy of Dede Olson)

TOP: *A curious young arctic fox watches researchers from the edge of its den.* (Photo by Sandra Lai) **BOTTOM:** *Many Arctic animals—including these arctic hare leverets resting next to Sandra Lai—show little fear of people.* (Photo by Charline Couchoux)

TOP: *Andrea Hazelton learns to read a topographic map at Inscription House Canyon, Arizona.* (Courtesy of Andrea Hazelton) **BOTTOM:** *Libby Ehlers tries out a daybed used by a large black bear in Yellowstone National Park.* (Photo by Matt Metz)

TOP: *Downed Enstrom helicopter at the bottom of Battle Canyon in northeastern Colorado* (Photo by Diane M. Thomas) **BOTTOM:** *A biologist holds a Rafinesque's big-eared bat captured during a bat population survey at St. Catherine Creek National Wildlife Refuge, Mississippi.* (Photo by Alison McCartney)

TOP: *Jan Hamber observes nesting California condors AC-2 and AC-3 (the Santa Barbara pair) in the Los Padres National Forest in 1982.* (Photo by Cindy Hawkins) **BOTTOM LEFT:** *Sophie Osborn carries a juvenile California condor in Vermilion Cliffs National Monument, Arizona, after testing it for lead poisoning.* (Courtesy of Sophie Osborn) **BOTTOM RIGHT:** *Marie Tosa fits a GPS collar and an ear tag on an immobilized white-tailed deer doe in southern Illinois.* (Courtesy of Marie Tosa)

Sophie Osborn radio tracks California condors at the South Rim of Grand Canyon National Park, Arizona. (Photo by Elisabeth Osborn)

TOP LEFT: *Maddie Halloran inserts a PIT tag in a juvenile salmon in Freshwater Creek, Humboldt County, California.* (Photo by Grace Easterbrook) **TOP RIGHT:** *Julia Smith carries adult male Mexican wolf 1342 to a processing crew during annual count and capture operations in Arizona.* (Photo copyright by Mark Henle, USA TODAY NETWORK via Imagn Images) **BOTTOM:** *Sloth conservation technicians enjoy a rare day off in Manuel Antonio, Costa Rica. Left to right are Holly Bradley, Masha Edmondson, Bruna Campos, Taylor McDowell, and Kate Derrick.* (Photo by Bruna Campos)

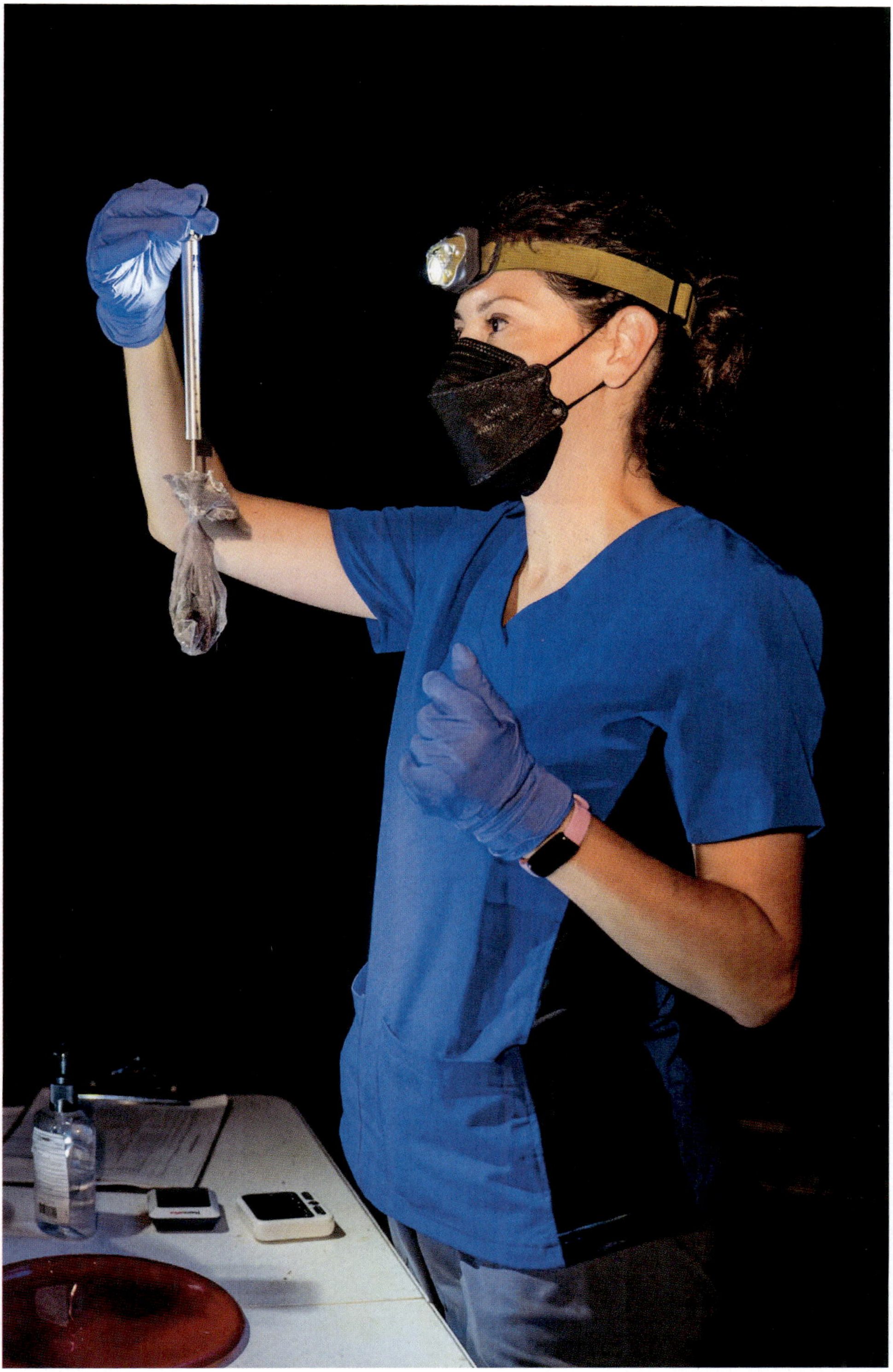

Lindsey Zarecky weighs a big brown bat in North Carolina. (Photo by Rebekah Robinson, Greensboro Science Center)

TOP: *Many women biologists juggle field work and motherhood in unique ways. Libby Ehlers, five months pregnant, measures vegetation with her son near Fortymile Dome, Alaska.* (Photo by Nicholas Ehlers) **BOTTOM:** *A curious yearling arctic wolf from the Eureka pack investigates a film crew's camp on Ellesmere Island, Nunavut, Canada.* (Photo by Kira Cassidy)

I stood there for a moment in silence, looking back toward the log where I had first seen the bear. In my mind's eye, I again saw the fuzzy brown head emerge from under the log as the bear looked up at me. I felt my breath quicken as I envisioned the bear launching out of its hidden bed. And once again I relived my terror at suddenly seeing a defensive bear charging right toward me.

Revisiting the site energized me. Experiencing the scene when it was safe to investigate it as a team, collect data from the elk carcass, and take the time to explore the area was immensely healing. After reexamining the layout of the meadow, visualizing the bear's actions, and replaying the events that had taken place here only two weeks earlier, I was able to associate the area with a positive, less traumatic experience. Almost immediately, my flashbacks disappeared.

Although getting attacked by a bear hasn't stopped me from enjoying bear country, where I've now worked on and off for more than fifteen years, the experience transformed me. Studying caribou confirmed for me that I'm now most comfortable in open terrain, where I can see long distances and in all directions. I'm also convinced that when the bear's saliva entered my bloodstream, I developed a sixth sense for detecting bears.

I have often felt a rush of anxiety and fear when hiking through closed-in forests, riparian areas, thickets, or areas known to have high densities of grizzly bears. Each time, my instincts have proven correct—alerting me a moment before surprising a sow and her two yearling cubs in Yellowstone, flushing a large sow with two young cubs in British Columbia, and encountering four grizzlies within a two-mile stretch while hiking with my family in Alaska. Each bear encounter has been thrilling but also nerve-racking.

Despite my unease, I still love to explore and share these wild places with my husband, Nick, and our two children. And I hope they will always treasure, as I do, the unpredictability of landscapes where just another walk in the park could, at any moment, turn into an adventure that forever changes their lives.

Nicholas Ehlers

Libby Ehlers is an applied ecologist with more than two decades of experience conserving endangered species across montane, boreal, arctic, and marine landscapes. Her lifelong commitment to understanding how humans impact ecosystems is the basis of her work to build collaborative, science-informed conservation solutions. She specializes in spatial modeling and using remote technologies—such as GPS video collars, drones, and remote sensing—to study wide-ranging species. Her PhD research at the University of Montana focused on summer foraging ecology of migratory caribou in Alaska and the Yukon.

Libby is a senior environmental scientist and wolf coordinator with the California Department of Fish and Wildlife. Living in the northern Sierra Nevada while keeping strong roots in western Montana, she enjoys music, canoeing, swimming, ice skating, camping, playing cards, cooking, and reading with her husband and their two adventurous children.

The Scream That Echoed for Decades

Susan B. Adams

Adjusting to fieldwork in the southeastern US was neither fast nor easy for me. I grew up in Northern California, attended college in Minnesota, and spent the next fourteen years working, playing, and studying in Idaho and Montana. None of this prepared me well for life as a fish biologist in the Southeast. My first summer of fieldwork after moving to Mississippi introduced me to many novel threats, like chiggers, fire ants, and cottonmouths—venomous snakes that are often abundant and well camouflaged in and around water. As my colleagues and I ate lunch on our truck tailgates near a river one day, we debated whether it was better to work where a grizzly might stalk and kill you for your tuna sandwich or where a serpent could slither toward you, unseen in the turbid water, and deliver a deadly bite before you even knew it was there. I voted for taking my chances with the grizzly any day—better the devil you know than the devil you don't.

I have a good scream. I like to think that it's bloodcurdling. I've used it to full effect on only three occasions—once intentionally and twice spontaneously. The first time was for a laugh. During a three-day campout on the California coast with my high school science club, my friends—a girl and two boys—and I decided to explore an abandoned mine shaft. As we approached the dark opening, we got to talking about scary scenarios we might encounter inside. We did not have flashlights, and I took the lead as we headed through the cave's twilight zone into the damp blackness. Once we were in Alcatraz-solitary-confinement-like darkness and I was a good way ahead, I let my scream rip. The boys were out of that cave faster than a bat out of hell.

The second time I screamed was out of real fear when I discovered a man lurking right outside my open window one brutally hot summer night in Idaho. I suspected by how fast he ran that he was more scared than I was after that scream. My neighbors heard it across our two large yards and through their closed windows, and called to see if everything was okay.

The third time I let loose my bone-chilling scream was in an Alabama stream during my first southern summer. I was doing freshwater mussel and stream habitat surveys with a group of biologists on Shoal Creek in Alabama's Talladega National Forest. Although fragmented by dams, the long riverine sections still retained many mussel species, including the finelined pocketbook and southern pigtoe, both listed under the Endangered Species Act. We were estimating mussel population sizes in relation to land uses to inform conservation efforts for this unique aquatic community. On the first day, we split into teams. Kevin and I would be assessing a couple of miles of stream habitat. We turned off the main forest road, opened a gate, drove a long way down a seldom-used two-track road that wound through forest and grassland, and parked near the study stream.

We began by walking downstream to our starting point, then spent the day evaluating the aquatic habitat as we worked our way back upstream toward the truck. This was the first time I had ever wet waded for work, meaning that instead of wearing chest waders, I walked in the stream wearing jeans and wading boots with felt soles that made walking on slippery rocks less treacherous. Previously, I'd always needed to wear waders because I was either working in very cold water or using a backpack electrofisher, which sent electricity through the water to stun fish. But it was a hot and humid summer day, the water meandering down the stream was tepid, and we weren't electrofishing. Wet wading was a nice change of pace, and the stream reflected the surrounding greenery while burbling a cheerful tune.

Although the scene was beautiful, it felt disconcertingly foreign to me, worlds away from my previous workplaces in western mountains, where scattered conifers, cottonwoods, willows, and alders grew along creeks with crystal clear water. This new environment, bursting with more shades of green than I had known existed and pulsating with animal life, left me feeling slightly claustrophobic and contemplating lurking dangers. Large deciduous trees crowded the banks, casting deep shade wherever their arching branches met over the stream. In brighter areas, the sunlight illuminated air so steamy it was visible. Poison ivy vines as thick as my wrist roped their way up trees and along logs that lay across the creek. The water was fairly clear, unlike the Mississippi streams near my new home, but it was still inhabited by venomous snakes. And although I generally liked snakes, I hadn't yet become accustomed to—or made my peace with—sharing water bodies with venomous ones.

The work went well, though by late afternoon I was getting worn out. I wilted in the oppressive humid heat that was still so alien to me. As the sun sank low in the sky, the shade under those big trees deepened, morphing into a murkier, more sinister-feeling darkness. Kevin had waded ahead, almost disappearing from my sight in the dim light. Frogs, toads, cicadas, and a host of unfamiliar creatures filled the air with their insistent calls. Piercing through that cacophony was the eerie *plop . . . splash . . . plop* of snakes dropping off branches into the creek as Kevin startled them up ahead. I knew that they were mostly nonvenomous water snakes, but still, every splash reminded me that I was walking through a stream full of serpents that had just been disturbed from their slumbers.

As I trudged along, my thoughts drifted through a series of increasingly dark scenarios. Soon, I was deeply immersed in a daydream about the leech scene in the movie *African Queen*—the one where Katherine Hepburn is in the boat, and Humphrey Bogart is pulling it through chest-deep water when he suddenly realizes he's covered in gigantic leeches. I often can't recall much about movies, but those leeches and Bogart's palpable terror upon discovering them were impressed upon my memory. Now I was reliving the scene in my head. So when a huge, dark snake suddenly reared out of the creek and lunged right at my face, I let out a horror-film-worthy scream unlike anything poor Kevin had ever heard in his life. Spinning around, he saw an enormous black snake attacking me.

He recounted later that his panicked brain couldn't decide whether to run like hell, leaving me to my grisly fate, or to rush back downstream to save me from my attacker. Heart racing, I confronted the lunging serpent, then . . . doubled over laughing as I realized I had stepped on the end of a long, thick stick, causing the other end, with its remarkably snakelike head, to rear up at my face. Even funnier was Kevin's terrified reaction.

As Kevin tells it (and he's told it often), that was the end of our story. But more than twenty years later, I still occasionally meet people at conferences who, as they start figuring out who I am and where I work, say, "Ohhh, are you the screamer?"

"Haha, yes, yes I am," I reply, "and did Kevin tell you the rest of the story?"

"No," they invariably respond.

It was just about dark when we finally got back to the truck. Kevin reached into his pocket to retrieve the keys—and couldn't find them. No worries, that often happens during fieldwork. He checked and rechecked pockets. We removed everything from packs. No keys. With a dawning sense of doom, he figured that when he'd visited the bushes at lunchtime, the keys must have fallen out of his pants pocket. Well, that was awkward. Not only could we not drive anywhere, but also our luggage was locked in the truck, and we didn't know if the other teams had driven back to town as planned or if they would come looking for us. We had neither cell phone nor radio.

Deciding our odds of rescue would be better on the main road, we started walking. As we headed up the two-track road under a huge yellow Alabama moon, we heard a loud roaring. Seconds later, a giant military transport plane from the nearby base flew slowly over us just above the treetops. For the first time in my life, I understood viscerally how vulnerable and terrified soldiers must feel, even in a thick jungle, when enemy aircraft fly overhead. We finally reached the main road and were walking along it when we heard a vehicle approaching. My ever-active imagination debated whether we were saved or should immediately hide because the psychos from *Deliverance* were about to descend upon us. Suddenly, lights rounded the corner . . . and after a brief deer-in-the-headlights moment, I was flooded with relief upon seeing that our colleagues had come to our rescue.

When we arrived at the motel, I had nothing with me but the clothes I'd been wearing all day—wading boots, soaked jeans, and a half wet, fully dirty shirt, which made for a most unusual night. The next day, Kevin's coworker drove several hours to bring him a spare key and help him retrieve the truck—and our bags. And although more drama was to come that week, including a close call with a real venomous snake, nothing was as memorable—or echoed for as many decades—as one tired biologist's high-decibel reaction to a violently lunging stick-snake.

Broken Wings

Diane M. Thomas

The rotors *whop-whop-whopped* overhead. Swaying gently from side to side, the Enstrom 280FX two-seater helicopter lifted smoothly off the airport tarmac and leaned into the wind, gaining altitude as it headed south to the Colorado rimrock country. I was en route to conduct an aerial search for raptor nests at a proposed wind farm location. Because raptors, which are protected by federal law, are vulnerable to being struck by fast-moving wind turbine blades, biologists evaluate raptor use of proposed project areas so that turbines can be sited where they will cause the least harm.

We had just made a pit stop at a small-town airport to refuel the Enstrom and grab the only food available—a couple of bags of salted peanuts from a countertop display and two cans of Diet Coke from a vending machine—and we were laughing at the paltry menu. Before takeoff, we had buckled our four-point over-the-shoulder seat belts, closed the doors securely, and donned our headsets, adjusting them to voice-activation mode. Headsets provide essential hearing protection in the noisy cockpit, which can exceed ninety decibels, and the voice-activated microphones allow clear, hands-free communication between the pilot and the passenger.

As I'd opened a file on my handheld GPS unit and checked my camera, maps, and data sheets, Craig had begun the multistep process of checking all systems, bringing the rotors up to speed, and preparing for takeoff. Like most pilots I've flown with, he was a highly skilled and experienced military veteran whose love for flying was reflected on his face. Now, as we gained altitude, he glanced at me, shrugged, and grinned.

Yesterday was the first time Craig had ever piloted a raptor nest survey, and he was enthralled with the amazing aerial views of golden eagles, prairie falcons, great horned owls, red-tailed and ferruginous hawks, and American kestrels sitting tight on their nests. From the helicopter, we also had a rare view of the eggs and the fluffy, gawking young that were visible in the nests when the adults were out hunting. Craig had enjoyed it so much that he had spoken excitedly about expanding his helicopter service to conduct more wildlife work. As he'd ferried me back to the airport, we had chatted, and I'd learned that he had endured a hellish time of late, suffering through an unwanted divorce, the loss of his father, and a cancer diagnosis, all within the past several months. Now he was hoping for a better future and relishing an opportunity to dream.

The bird's-eye view of landscapes, as seen from a plane or helicopter, always enchants me. I never tire of reveling in a panorama to which human eyes were not privy until sometime after the advent of "flying machines" in 1903, when the Wright Brothers' first powered flight—a twelve-second, 120-foot venture, 20 feet above the ground—severed humankind's tethers to terra firma. Flying transports me to another dimension, allows me moments of reflection to imagine life as a feathered denizen of the air, and resets my view of the world and my infinitesimal place in it.

The first time I conducted a raptor nest survey by helicopter, I was so excited (and uninitiated) that I used my binoculars freely and within an hour was so overcome with motion sickness that I had to sit out and recover for several hours. Hundreds of flight hours later, I still occasionally succumb to queasiness even though I now employ tricks such as watching the horizon to reorient my brain and inner ear to where I am in space. Long, hot summer days are particularly challenging when I'm perched within a small plexiglass bubble while the rotor blades strobe a shadow-flicker through the piercing overhead sun and pound out a tempo that's audible despite my headphones. The occasional smell of fumes wafting into the cockpit through the tiny air vents, especially right after refueling, only adds to the nausea at those times. These days, if I am planning on a full-day flight, I bring an airsickness remedy. I want nothing to diminish my joy when conducting aerial surveys. I also want no distractions, because surveys of this type demand constant vigilance by both pilot and biologist.

Aerial raptor nest surveys are, to me, a bit like speed-reading the pages of a densely packed book, looking for a specific word or phrase. My eyes scan rapidly, reading the cliff left to right, right to left, up and down, layer by layer as I search for stick nests, as well as cavities and ledges that are painted with a raptor's telltale white excrement. The surveys also are like delightful scavenger hunts that fill me with anticipation and excitement as I seek the treasures displayed on cliff ledges or tucked into crevices. During these surveys, minimizing disturbance to nesting raptors is paramount, so I always instruct the pilot to maintain altitude and distance from active nests and to move quickly through the birds' territories. Though it may seem counterintuitive, ground-based surveys, especially if the biologist is on foot, generally require more time spent in the nest vicinity and can be more stressful for the birds.

Teaming up with an exceptional pilot is essential for conducting successful surveys. Although a second pair of eyes can be helpful for locating nests, a pilot needs to watch diligently for dangers such as power lines, trees, other aircraft, and birds, any of which could entangle the rotors and send us plunging earthward. Through the years, nesting ospreys have attacked my survey helicopters on several occasions, and without evasive maneuvers by one or both parties, these games of chicken could have been deadly for all. Raptor surveys in a fixed-wing aircraft can be equally perilous. A pilot I know described the terrifying moment when he looked out of his cockpit window and saw a bald eagle hurtle between the wing strut and the body of his small high-wing plane—a triangular space smaller than an eagle's full wingspan. An eagle collision with the plane's propeller or windshield would have spelled certain disaster.

Piloting a helicopter is a full-body endeavor. In addition to the 360-degree spatial awareness such flying demands, one hand must constantly guide the cyclic pitch control, which is like a joystick protruding from the floor that controls the direction of the aircraft (forward, backward, left, and right) by moving the individual rotor blades. The other hand controls the collective pitch (which moves all the blades simultaneously) and the throttle, which regulates rotor speed and engine rpms. Meanwhile, both feet are engaged on the anti-torque pedals that control the tail rotor, which counteracts the torque effect of the main rotor and controls the heading of the helicopter while hovering or making hovering turns.

As if all that were not enough to occupy a person's attention, the pilots I have flown with also are amazingly adept at reading my body language to determine where I need the aircraft to go. After a short time, it is as if the pilot and I are of one mind when it comes to where to fly and how to position the aircraft so I can view the raptor nests, and it takes very little on my part to elicit a change in direction or indicate where I want to go next.

Aerial raptor nest surveys also demand a great deal from the helicopter, especially in the West's high-altitude, complex terrain. Flying alongside cliffs and up and down narrow side canyons, skimming the treetops, navigating the air currents, and hovering while a biologist photographs nests and obtains their locations with a GPS all contribute to the challenge. As any hummingbird might confirm by its near-constant need to feed on high-energy nectar, hovering is by far the most energetically demanding type of flight and requires that a pilot execute a delicate balance among all three primary flight controls.

A helicopter's performance is reduced with increasing density altitude (a function of altitude, ambient temperature and humidity, and atmospheric pressure), wind speed, and weight. Given that density altitude and wind speed cannot be controlled, the third factor—weight—is carefully monitored and regulated by limiting the number of people onboard and by not filling the fuel tanks to maximum capacity in especially demanding conditions. Since helicopters may cost more than two thousand dollars per hour to rent, the pilot must balance the fuel load needed to maximize survey time and avoid unnecessary side trips to refuel with ensuring that the helicopter is light enough to safely navigate current flying conditions.

As Craig and I approached the line of cliffs and box canyons (narrow, steep-walled canyons with only one entry or exit), we moved in close to the sheer rock walls, resuming our nest search where we had left off before refueling. It was a pleasantly warm spring day, and we had a full fuel tank, drinks, and snacks. We would be able to fly for several hours before we had to return to the airport to refuel.

I had barely begun to scan the cliffs when a large inactive golden eagle nest loomed on the ledge in front of us. Within their breeding territories, raptors often build and maintain alternate nests, which they may use in subsequent years. Recording data on the location and condition of these inactive nests is useful for determining a raptor pair's nesting territory boundaries. Since disturbing a bird isn't a concern at inactive nests, the data needn't be recorded as hastily, and hover times are slightly longer.

I pointed upward to request that Craig rise thirty feet higher so that we could hover above the nest to obtain an accurate GPS location. I then quickly wrote a few shorthand notes on my legal pad describing the nest's condition and grabbed my camera for a photograph. As Craig started to pull away, I held up my hand to request a bit more time to snap a picture. The Enstrom tilted slightly and began to lose altitude, and Craig quickly shook his head, indicating that he could not hold the hover. I nodded, but we continued to drop toward the narrow canyon bottom.

"We're going down." Craig's voice was calm in my headphones. I looked over at him and grinned. Surely he was joking. But as the trees in the canyon and the rock walls of the cliff rose to meet us, I realized he was deadly serious.

I have been in grave situations beyond my control several times in my life. Each time, I've felt a calm, almost peaceful resignation to fate. More than twenty years earlier, I had accompanied a biology professor and a group of college students on a trip to Africa to assess the feasibility of an ecotourism/conservation venture in the Zanzibar Archipelago. Our hydrofoil boat trip from Zanzibar to Pemba Island had been an exercise in seasickness as the notoriously turbulent waters of the Pemba Channel sickened one member of our party after another. Several weeks later, we opted to take a small charter jet back across the channel to Zanzibar to avoid the boat trip. While boarding the jet, we laughed at the two fire trucks parked in the tall grass off the edge of the tarmac. One was up on blocks, and both looked like they belonged in a parade of antique vehicles. The trip was to be a short one—approximately eighty miles airport to airport, including about thirty miles over the Pemba Channel.

Once airborne, I reached for my camera and started taking photographs, ecstatic to have such an expansive view of the islands. Midway across the choppy Pemba Channel, absorbed in the beautiful scene below, I suddenly realized that a sickening white mist was filling the cabin. Smoke quickly followed. The pilot tersely informed us that the hydraulic system that powered the wing flaps and landing gear was on fire and that we should prepare for a crash landing by bracing ourselves and putting our heads between our knees. Sneaking a last glance at the suddenly ominous sea below, I fervently hoped we could reach land before crashing, and I prayed for a safe outcome. We covered our noses and mouths with our shirts to filter out some of the noxious fumes from the smoke and burning hydraulic fluid.

The pilot quickly changed course to return to the Pemba airport and, in a proactive move, dropped the landing gear, hoping that enough hydraulic fluid remained in the system to lock the wheels into place so that they would not collapse on touchdown. If they collapsed, we'd experience a "gears-up" or belly landing, increasing the risk of injury and the potential for fire to engulf the aircraft. As we approached the airport, I risked a glimpse out the window and saw one of the fire trucks we had so recently laughed at being pushed out onto the tarmac in preparation for a fire or a crash landing—*our* crash landing. The ground rushed toward us. Our tires bumped the pavement. The plane bounced and the tires bumped again. And then a third time, as the landing gear held and we settled into a normal landing. The pilot's quick thinking and Providence had protected us from a potential catastrophe.

Now, riveted by the rapidly approaching trees and rocky ground, I knew that this time a crash landing was inevitable. And if we burst into flames, no fire truck—ancient or otherwise—would be there to help us. Once Craig realized that the Enstrom—loaded with fuel—did not have the power to climb out of the canyon because of the strong wind streaming down from above, he managed to keep the rotors from clipping the surrounding cliff faces as the wind flung the aircraft downward.

As we plummeted to the canyon floor, the stubby broken trunk of a leafless cottonwood smashed through the plexiglass alongside the pilot's door, shattering the window and slamming into Craig's thigh. In a rapid-fire one-two punch, one skid hit the ground at an angle and broke, and as the helicopter lurched drunkenly, the spinning rotors struck the ground forcefully and bent, yanking the aircraft forward. The second skid hit, then rolled sideways like a turned ankle, and the aircraft finally settled onto its side, accompanied by a roar of silence. The impact had shattered the plexiglass bubble in front of me, and my clipboard and GPS unit had been ejected, landing on the ground more than fifteen feet away.

Craig and I immediately performed silent assessments of our own and each other's physical condition. No broken bones. Neither of us were bleeding. Both conscious and aware.

Woodenly, we unsnapped our seat belts (never was I so happy to be well strapped in) and crawled out of the wreckage. A hole in Craig's flight suit marked where the tree had struck his thigh, but fortunately he was only deeply bruised rather than seriously injured. Later, however, I learned that he had also injured his wrist in the impact.

We moved away from the aircraft warily, scanning the wreckage for any fuel leaks that could lead to a fire. No drips, no odor—good news. Next, my eyes roamed the canyon walls, trying to find a climbable route to the top since we had no cell phone service at the bottom and the helicopter radio was damaged. Several routes looked possible, but as we stiffly began the ascent, we found they were choked with dry, prickly tumbleweeds that concealed the routes' true depth and steepness. After several false starts, we found a challenging but doable route. Shakily, we clawed our way up, finally scrambling up the steepest rock outcrop to crest the cliff's rim. We found ourselves on flat land, approximately 150 feet above the fallen helicopter.

Now the reality of our plight and the challenges still facing us sank in. We were in a remote area, and though we thankfully now had cell coverage, both of our phones had perilously low batteries. The first order of business was for Craig to notify the National Transportation Safety Board (NTSB) and his insurance company. We had to call the sheriff's department to report the crash and obtain transportation back to my truck at the airport. We had to notify our bosses and loved ones and call the airport where the helicopter was slated to be hangared that night to tell them we wouldn't need the space. Craig also wanted to climb back down and retrieve the radio from the wreck to salvage what he could of the expensive equipment.

But despite all we had to do, neither of us moved. Instead, for nearly an hour we sat in solitary reflection, gathering our wits, giving thanks for our safety, and reconstructing what had happened so that Craig could provide the NTSB with a comprehensive and coherent description of the crash. By the time we called the sheriff's department, it was around 2:30 p.m.

The crash site (which we later learned was named Battle Canyon—an apt moniker given our harrowing incident) was difficult to approach by vehicle. Not only was it on private land far from a public road, but also there was no address or easy way to direct ground vehicles to it. Fortunately, my GPS unit was still working and had enough battery charge for me to obtain our location's latitude and longitude. By the time the sheriff's office rounded up and dispatched a small plane to fly over the area and find us, and then provided deputies with directions for driving overland to our crash site, it was nearing 5 p.m. The March sun was approaching the horizon and losing its warmth. Emotional and physical exhaustion had seeped in, sequestering us in the confines of our own silent musings.

As we waited for rescue, we heard the droning of an aircraft and then watched as a small red-and-white plane flew over, making several passes and dipping its wings to signal that the pilot had seen us. Soon after, we spotted dust plumes in the distance, and in short order four sheriff black-and-whites came jolting across the top of the mesa. After exchanging introductions and formalities, we all walked to the cliff edge to view the downed helicopter.

After a reverent silence, one of the deputies muttered, "You were lucky to walk away from that one, let alone climb out of the canyon afterward."

"I'll think twice about flying drug reconnaissance for the sheriff's department going forward," a second deputy soberly chimed in.

The fire chief shook his head and marveled, "Would have been a much different scenario had that full fuel tank ignited."

As we wearily piled into a squad car for the forty-five-minute drive to town, all my remaining energy drained out. Closing my eyes, I tried to relax away the raging headache and sore muscles that had started to nag at me in the aftermath of the crash. At the sheriff's department, we endured an hour or so of filling out paperwork, answering questions, making phone calls, and washing down Tylenol with a couple of glasses of water before a deputy finally loaded us up and drove us across town to the airport, where my company pickup was parked. It was a somber arrival. I could only imagine what Craig must have felt coming back to the airport by ground transport—like a cowboy stripped of his horse, clomping about the prairie in boots that were never meant for walking. We had arranged that I would drive him halfway to Denver, where his friend would pick him up and take him the rest of the way home.

On the drive, between long moments of thoughtful silence, we spoke of his future and offered a prayer of gratitude that we had escaped a tragic outcome. But even though we had survived, I couldn't help but think about the joyous abandon with which Craig had discussed his dreams for the future that very morning, and I sadly contemplated the dark shadow that had so suddenly obscured them.

Flying is one of the most dangerous tasks wildlife biologists undertake. Inclement weather, loss of power, and mechanical problems are ever-present dangers. Additionally, power lines, tree branches, cliffs, and birds pose collision risks to aircraft. All these hazards are magnified during biological surveys, which generally are conducted at low altitudes and often in rugged terrain.

I have known two biologists who went down while working. One survived a hard landing in a helicopter in Texas. The other went down in a small plane while radio tracking polar bears near Point Barrow, Alaska. Despite exhaustive searches, neither the plane nor the two biologists and pilot were ever found. I also know of a helicopter crash that killed a pilot and two fisheries biologists who were conducting salmon surveys in the Pacific Northwest.

Although my nest survey had been interrupted by our crash, my company was still contracted to complete the work. Oddly, the first available pilot I could find had trained Craig in the specifics of the Enstrom years earlier. Ted and I arranged to return to Battle Canyon and resume the surveys within the week. I was unsure how I would feel once airborne. I did not want my joy of flying to be stripped away by the crash, and though I felt ready to return to the air, I wondered if seeing the crash site and the broken helicopter below would elicit a visceral fear response.

On the morning of the flight, I met Ted on the tarmac near his Bell 206 Jet Ranger. The helicopter was nearly three times as expensive to rent and much larger and louder than the Enstrom, making potential disturbance to nesting raptors a bigger concern. However, it was also much more powerful, with a turbine-powered engine rather than a piston engine. We took off, flying north toward Battle Canyon, where the Enstrom still lay on its side awaiting decisions on how best to retrieve the wreckage. Up in the air, so far, so good.

As we neared the site, I saw the crumpled Enstrom below us, fallen like a dragonfly stricken in midflight. Suddenly, the Jet Ranger dropped precipitously, and I felt the sinking sensation that I had experienced during the crash. Were we going down? Again?

I quickly glanced at Ted, searching his face for any signs of trouble. With an inscrutable expression, he looked back at me, then pulled up out of the descent and began to instruct me on helicopter safety. Never ever use the wrong aircraft for the job, he lectured. This was a potential wind farm location, with the attendant wind speeds one would expect. The survey required a lot of hovering in high-altitude, dissected topography with unpredictable and sometimes turbulent winds blowing up and down the box canyons. While Ted made clear that the Enstrom was an excellent helicopter, he argued that it had not been the right one for this job. Furthermore, he insisted that we should never have taken on a full load of fuel. He went on to expound on the merits of turbine-powered versus piston engines. Piston engines did not come close to having the power of turbine engines and were six times more likely to fail midair than were turbine engines. Lesson over, we moved on to the survey and completed the job.

I suspect that Ted imitated the loss of power and the drop in altitude that Craig and I had experienced in the canyon above the crash site to impress upon me the importance of what he was telling me. Perhaps he also wanted to assess whether the crash had transformed my love of flying into a fear of flying. At the time, though, his actions felt calculated and cruel. Nevertheless, having been tested, I was relieved to have passed muster. I still love to fly and have conducted many more fixed-wing and helicopter surveys since that day.

Through the years, both before and after the crash, I have often heard pilots remark somewhat casually, "There are two kinds of helicopter pilots—those who have crashed and those who will." With the significant increases in helicopter safety features and pilot training in recent decades, this phrase may no longer accurately reflect the risk associated with helicopter flight; however, it serves as a reminder of the inherent risks involved in conducting aerial surveys. Each time we wildlife biologists fly, our cumulative odds of crashing probably also increase. But given that I've now weathered several potential disasters, I'm hoping there'll be no more broken wings in my future.

Courtesy of Diane Thomas

Diane Thomas grew up in Nebraska, where her mom filled her four-year-old hand with salt and told her that if she could sprinkle it on a bird's tail, the bird would let her catch it. This effort, though unsuccessful, kept her occupied the entire summer and sparked her lifelong love of birds. (Thanks, Mom!) She earned a bachelor's degree in English and art and five years later moved west to earn a bachelor's degree in wildlife conservation and a master's in zoology and physiology at the University of Wyoming.

Diane's favorite fieldwork as a wildlife biologist involved avian studies, particularly raptor nest monitoring. She also volunteered for four years at the Laramie Raptor Refuge, where she fell in love with a sweet, nonreleasable short-eared owl and many other injured raptors. Diane lives in Wyoming with her beloved Australian cattle dog, Lark, and loves visiting friends and family, birding, reading, gardening, hiking, kayaking, and exploring new places in the Wobbling Vireo, her trusty camping van. She enjoys international travel and experiencing new and exotic birds, wildlife, flora, cultures, and cuisines.

Always Carry a Stick

Brianna Mann

Having a great passion for field biology can help a person overcome many of the struggles faced during fieldwork—exhaustion, discomfort, hunger—but some days in the field are bound to test even the most devoted biologist. One of these days occurred during the July I worked on a field crew studying the endangered New Mexico meadow jumping mouse in northern Arizona and New Mexico. We were spending a couple of weeks in the Gila National Forest setting track plates, ingenious contraptions designed to help us find jumping mice—if there were any to be found—by documenting their tracks.

The track plates consisted of a rectangular plastic container, about the size of a shoe box, turned upside down with paper and an ink pad attached to the inside of the lid, which formed the bottom plate. Mice entered the track plates through one of the holes we'd cut into each end of the containers, and when they walked to the center to feed on the bird seed and peanut bait, they'd sign the guest book by stepping on the ink pad and then leaving their distinctive long-toed tracks on the paper.

A few days into the project, we'd found a whole lot of nothing. The habitat we were surveying lacked the lush carpets of herbaceous plants—forbs, sedges, and grasses—that the mouse needs. The area had been scoured by floods that left sparse vegetation in their wakes. Undeterred, we continued to set track plates in an area bordering a shallow, gently flowing stretch of river no more than ten feet across and a couple of feet deep.

It was monsoon season in the Southwest. One morning we drove for hours to the study site in the pouring rain, only to arrive and watch dejectedly as the water rose steadily and our precious track plates sank slowly into the mud. One look at the swollen river told us that even if we had captured any evidence of jumping mice on our track plates overnight, the mouse tracks had now been wiped away by the current. But even with no retrievable data, our remaining track plates were at risk of floating away. Some of the plates were already drifting downriver, lost to the flash flood. There was only one way to save them.

"All right. Get as many as you can, and if the water gets too high, get out of there!" our field crew leader shouted over the storm. We pulled on our waders and backpacks and prepared to hustle. "And don't get electrocuted!" she added, as lightning ricocheted across the sky.

We sprang into action as we had a thousand times before—only this time wading through a slurry of unusually fast-flowing water and mud, snatching up track plates as quickly as we could. We had set more than two hundred along both sides of this river section, a typical transect, and our crew of five started retrieving them at each end, working fast and watching out for our teammates.

The rain fell harder and the water rose higher. I kept one eye on the rising water and the other on my teammate, Matt, who was on the other side of the river. I barely glanced at the track plates as muscle memory helped me snatch up, disassemble, and swiftly stash each one before moving on to the next. We were a well-oiled machine.

I stood in knee-deep water on what was usually the edge of the river. The storm roared so loudly that I could hardly hear Matt. The river was now a rushing fifteen-foot-wide torrent. We shouted back and forth about the lost causes—those track plates that had already been swept downstream and could not be saved except by putting ourselves in danger. Reaching down to grab yet another track plate, I lifted the top off as I had done countless times before.

Suddenly, time slowed. A western black-tailed rattlesnake was coiled on the ink pad taking refuge from the storm. We saw each other at the exact same moment and both reacted in what felt like the same slow motion.

Just as the snake recovered from its surprise and reared to strike, I leaped four feet backward as though my boots had suddenly developed metal springs. The timing of the snake's strike and my narrow escape was perfectly synchronized. My heart pounded for a moment as I processed what had just happened. Matt spun toward me, taking in the scene. The waters continued to rise, but for a moment, Matt, the snake, and I were frozen in place. Then, glancing away from the snake, Matt and I caught each other's eye.

"Yep, time to go!" we agreed, and tore ourselves away from the mayhem as the flood escalated. Drenched and muddy, we rushed uphill to the safety of the nearby Forest Service road, where we relayed my jarring encounter to the rest of our crew.

"So, new addition to our protocol," our field leader promptly proclaimed. "The next time we check track plates, everyone will first hit the container with a stick."

I learned many invaluable lessons the summer I searched for jumping mice, but above all I discovered that when doing fieldwork, it's a good idea to go with the flow. And you might just have to make up new rules as you go.

Conservation Under Fire

Gina Piazza

When I envisioned life as a wildlife biologist, I pictured days spent outside under blue skies, working blissfully with wildlife in idyllic settings. People, politics, and being covered in mud were not part of that picture. But I've learned since then that wildlife and people are inextricably linked, and working as a field biologist requires navigating humanity and its complexities, no matter where you find yourself.

My first lesson began with a black-spotted, reddish frog not much larger than a ping-pong ball. Oregon spotted frogs are listed as a threatened species under the Endangered Species Act. They once inhabited lakes, ponds, wetlands, and riverine floodplains all along the West Coast from Canada to California. But due to the introduction of nonnative plants and animals, habitat loss, pollution, and other stressors, the frog now lives in less than 21 percent of its former range.

With populations dwindling, biologists developed a recovery plan that included a pilot reintroduction project for this important member of Pacific Northwest wetland ecosystems. Reintroducing captive-reared animals to the wild is a complex process. To raise frogs in captivity, egg masses were carefully collected from the wild in early spring, and the resultant tadpoles were fostered at several zoos and a correctional facility until they metamorphosed into frogs. (I'm not sure why, but it seemed like the frogs nurtured at the correctional facility were always the largest and plumpest!) Before being released into the wild in the fall, each frog had an identifying chip—known as a PIT (passive integrated transponder) tag—inserted under the skin on its back. Each tag contained a number indicating where the frog was reared and when it was released.

To establish a self-sustaining population, about five thousand frogs were released over a five-year period in a carefully selected, pristine wetland on the Pacific Northwest's largest military base. (Nationwide, thousands of acres of nearly untouched ecosystems on military bases provide habitat to many species of concern.) My job was to locate as many frogs as I could throughout the seasons after their release into this wetland.

The frogs had dispersed and could have a home range of five acres or more, so one of the ways I found them was by using a PIT tag monitor. The monitor consisted of a wand that looked like a metal detector but had an extra-large battery carried as a backpack. The device beeped when the wand came within a few feet of a PIT tag. Upon hearing a beep, I tried to find the frog, either by flushing it out of the vegetation or by digging a few inches into the mud, to determine if it was still alive. Detecting the frogs and determining how many had survived would help us understand whether the reintroduction methods could successfully establish a sustainable population on the military base.

For months, colleagues and I donned chest-high rubber waders and systematically hiked the six expansive, interconnected wetlands surrounding the release site, searching for these small, threatened creatures. Weeks would go by without any detections. During those times, it was hard not to feel like we were trying to find a needle in a haystack. However, we dutifully continued to climb over submerged logs in hip-deep water and muck during all sorts of inclement weather, while carefully protecting the heavy, expensive, and fragile monitoring equipment.

Despite undergoing initial background checks by the military and passing multiple ID checkpoints to enter the base, once we were in the tranquil preapproved study area it was very easy to forget we were on an active military base. We never saw any people except our other team members. One notable day, Bonnie (a fellow biologist) and I decided it was time to think outside the box and search for frogs beyond the prescribed wetlands. We figured that the frogs didn't know to stay within the boundaries drawn on our map, so perhaps we should venture farther, too. I was always excited to work with Bonnie, who was full of ideas, had a lot of stories, and was always in the middle of multiple projects. With dirt that seemed to be permanently embedded under her nails and twigs often stuck in her hair, she was definitely a kindred spirit. We both also tended to follow our instincts and stray off the beaten path. We figured the frogs might do the same.

Accordingly, we studied satellite images of the base, looking for other wetlands where the frogs might congregate. To our delight, we found another wetland a bit farther from the release site but still connected to the same creek. Even though no road led to this newly identified location, we were up for the challenge and excited to explore unknown lands.

Early in the morning, armed with coffee and a new sense of optimism, we drove for about an hour, jolting along on an old, rutted logging road that led deeper into the military base. Our progress was blocked sooner than we expected by some downed trees. Undeterred, we unloaded our kayak, carried it and all our gear about half a mile, and launched from a misty creek bank.

We floated peacefully, letting the slow current take us down the stream—occasionally navigating around fallen logs and beaver dams—and enjoying the serene water on a quiet morning. The surface was as smooth and clear as glass, and we could see fish swimming leisurely below us. A majestic great blue heron walked stealthily along the foggy bank, patiently searching for its breakfast. I felt like an early explorer as a family of curious river otters followed our kayak for part of the journey. We paddled past groups of elegant, colorful wood ducks, and after about an hour of thinking *I can't believe I get paid to do this*, we checked our GPS unit and confirmed that we had arrived near the new wetland area. We tied the kayak to a bundle of reed canary grass and then hiked for another mile or two into what felt like the middle of nowhere.

Excitedly, we finally began our search for frogs, focusing on the ground and the interface between the water and the land. The frogs would be either in shallow areas along the margin of the stream or buried under a few inches of mud to stay warm.

We moved slowly along the water's edge, absorbed in our task, until a very unnatural whistling sound pierced the quiet. Then BOOM! The ground shook. We looked at each other in shock, trying to determine what was happening. Seconds later, another whistle and BOOM! Suddenly, machine gun fire erupted just on the other side of a row of trees less than thirty feet away, propelling us into action. Instinctively, we both threw ourselves facedown onto the mud, returning fire with a volley of expletives. Apparently, soldiers were training, oblivious to the presence of two muddy biologists on enemy land. Knowing that lying on the mud wouldn't keep us safe, we hastily agreed to run for the kayak.

Sprinting for safety, with bullets whizzing past us, I had only one thought—*ESCAPE*. Each time we heard another explosion or rattle of gunfire, we threw ourselves onto the ground again. Upon reaching our kayak, we leaped in and paddled like we were charging for the finish line. Flying out of the water, we thrashed through willow branches and thorny blackberries, dragging our kayak back to the truck. Feeling some relief once in the safety of our vehicle, but still rattled, out of breath, and a bit stunned, we called it a day.

Not surprisingly, we didn't find any frogs that day, but I learned a couple of unforgettable and important lessons. First, no matter how remote and wild my surroundings might seem, civilization is often lurking just behind the next tree. Second, and most important, while being bold and exploring new places inspires me as wildlife biologist, I learned that I *must* register for all areas that I plan to visit on a military base! Above all, I felt fortunate that this was the first—and I hoped the last—time I experienced what too many endure daily in war-torn countries.

Needless to say, Bonnie and I both meticulously followed the rules from that point on. There's nothing quite like getting shot at to make you a stickler for following protocols—and that's true even for a couple of intrepid biologists with twigs in their hair.

A Congregation of Alligators

Alison McCartney

The first time I met an alligator in the wild, I was canoeing through a cypress swamp in the heat of summer. I've been afraid of them ever since.

I was a young graduate student just starting fieldwork for my master's degree. My research in the St. Catherine Creek National Wildlife Refuge in southwest Mississippi focused on several colonies of two rare bat species: Rafinesque's big-eared bat and southeastern myotis. In addition to roosting in hollowed-out cypress trees, both bats also roost in old buildings. Several abandoned houses and old cisterns in or near the refuge contained colonies of both species. I regularly monitored bat numbers at each location and identified features that characterized the bats' roosting areas to better understand the types of abandoned structures they preferred.

The refuge is bordered on the west by the massive Mississippi River and on the south by one of its tributaries, the Homochitto River. St. Catherine Creek flows through the middle of the refuge. Because so much water moves through this fairly flat landscape, the majority of the refuge floods each year and consists of beautiful cypress swamps along with bottomland hardwood forests. Water seems to drip from the trees in these majestic forests. I worked several hours away—at the Mississippi Museum of Natural Science in Jackson—during the week and spent most of my weekends at the refuge collecting data for my graduate project.

Having always been very independent, I relished the weekends by myself, exploring the forty-thousand-acre refuge and trekking through knee-deep water in the middle of cypress swamps. It felt like my own private playground since I rarely ran into anyone else, and the freedom of being in those glorious, expansive forests was exhilarating. Because I was raised in Southern California, I found swamps new and fascinating. Curtains of Spanish moss draped from eighty-foot-tall cypress trees that were six feet wide at the base. The constant high hum of cicadas and the sunlight filtering through the canopy and reflecting off the swamp water created a setting that could only be found in the Deep South. Time seemed to slow down in this magical place, and I would get lost in the pure beauty of it.

While canoeing through one of these swamps one hot summer day to get to one of the abandoned houses, I had my first alligator encounter. For once I wasn't alone but rather was accompanied by a volunteer. The duckweed was so thick on the water's surface that we had not been making any progress using our paddles, so Jake was walking ahead and hauling the canoe through the knee-deep water with a rope. As he pulled the canoe, I suddenly heard rushing water to my left. Looking over, I saw a six-foot alligator racing through the water with its head raised, literally roaring. My heart stopped as the alligator headed straight for my volunteer.

"Gator!" I yelled. Jake, who wisely carried a walking stick, pounded the stick against the ground below the water several times. Maybe this wasn't his first run-in with an alligator because, to my surprise, the gator whipped around, smacked its tail hard against the water, and disappeared. Evidently the vibrations caused by pounding the stick on the ground were enough to make the alligator change course. Regardless, I've been terrified of alligators ever since.

My next alligator encounter occurred at the same refuge the following summer. Several volunteers and I were preparing to trap bats with fine-mesh mist nets, each twelve feet long by six feet tall, strung between two aluminum poles. We planned to place our four nets over a small pond in the relatively shallow water within fifteen feet of the bank. Bats often visit water sources for a drink before heading out on their foraging flights and will get entangled in the nearly invisible nets that intersect their flight path. By identifying the bats we extricated from the nets, I could determine which bat species lived in the refuge. I also outfitted some of the captured bats with small radio transmitters so I could track the nocturnal flyers to their roosts and better understand how they used their habitat.

As we set up our nets shortly before sunset, I realized that a three-foot-long alligator was living in the pond. It swam toward us until it was about thirty feet away and then watched us as we worked. Each time we finished setting up a net and went on to the next, it followed us to the next spot, stopped thirty feet away, and again silently watched. It was a little eerie, but I wasn't too worried about such a small alligator.

And then it got dark. We unfurled our nets and got busy catching bats. We caught several species that night, including two evening bats and three eastern red bats. Each time we went to check a net, I heard a commotion in the water about thirty feet away. In the beam of my headlamp, I saw hundreds of small silver fish jumping out of the water and, shortly thereafter, the eyeshine of that alligator. It followed us all night.

Whenever I checked the nets or worked to untangle trapped bats, I found myself glancing uneasily over my shoulder, searching out the alligator to make sure it wasn't coming closer to me. A knowledgeable friend once told me that alligators are more curious than dangerous, and this one was certainly too small to cause me any serious harm. Nevertheless, I was disconcerted to discover as we were closing down our nets at one in the morning that one of the nets had a three-foot hole in the bottom. I'm guessing that alligator had gone right through the net, and we hadn't even noticed!

Later that summer, my growing fear of alligators became all the more apparent. I was at the refuge with a fellow biologist who would one day be my husband. Before we started mist netting that night, Marshall was fishing in a deep drainage ditch while I sat on the bank in my camp chair, reading a book and enjoying some rare downtime. We had been there for about an hour when I suddenly realized there was an eight-foot alligator sunning on the opposite bank, so covered in mud I hadn't even noticed it. Perhaps the gator felt my gaze, because shortly after I spotted it, it slipped down the bank and into the water.

From that point on, even with my eyes focused on my book, my mind was consumed by the alligator. I was sure it would swim down the ditch, come out of the water unnoticed, circle behind me, and wait for its chance to attack. I soon convinced Marshall that it was time we left. My logical, prudent fear of and healthy respect for alligators was quickly becoming an irrational paranoia.

That's when my alligator nightmares began. I would dream that while I was netting a pond for bats, a ten-foot gator grabbed my leg and dragged me underwater to drown me before eating me. When I was younger, I had had nightmares about sharks attacking me, perhaps because I lived next to the ocean, but now sharks had evidently been replaced by gators. My worst alligator dream occurred many years later, after Marshall and I were married and had three kids. I dreamed we were hiking at the refuge when my two-year-old rounded a corner in front of us and was promptly attacked and eaten by a ten-foot alligator as we watched helplessly. The horror of it has haunted me ever since.

Four years after I finished my master's research, my unwanted encounters with alligators continued while I was conducting surveys at the Yazoo National Wildlife Refuge in the Mississippi Delta. Alligators were everywhere! Driving in the refuge one day, I turned the corner and had to brake to avoid hitting a massive twelve-foot gator basking in the road. My passenger, a refuge employee, told me that some kids had shot this poor gator in the eye a couple of years earlier. As he told me the story, the one-eyed gator slowly slipped down the bank and into the water. Pity for this unfortunate giant began to soften my fear of gators—slightly.

But my abject fear returned a few years later when I was looking for potential sites for mist netting bats in Mississippi's Bienville National Forest. I was scouting in the woods alone when I came upon a swampy area adjacent to a wide creek. The banks were fairly steep, and due to the murky water, I couldn't tell how deep the creek was. Standing on the bank, I used a stick to test the water depth to determine whether it was shallow enough to set up our nets. A light drizzle was falling and quickly turned the dirt bank into a clay waterslide. As I leaned forward, my feet slipped out from under me and I slid down the muddy five-foot bank. I plunged into water over my head and did not touch the bottom, then came up sputtering, shocked at the unexpected turn of events.

Alligators were my first concern. Although freaked out, I tried to calm myself down. I had almost managed to get my breathing back to a somewhat normal rate when I turned and saw the imposing streambank that was my exit route. Picturing a slew of gators on the water's surface eyeing my back and slowly moving in, I frantically attempted to scramble up the steep, muddy bank but slipped back into the water. Panic rose as I again tried to claw my way up the bank, struggling to gain a foothold. After what felt like an eternity, I finally made it. I sat there for a minute in the rain, soaking wet, staring at the water, still expecting to see hungry alligators closing in.

I knew I was being ridiculous, but my heart continued to pound. Jumping to my feet, I walked quickly through the woods toward my truck, which was about half a mile away, trying to put aside my irrational fear. Consumed by my anxiety, I belatedly became aware of my surroundings and realized I had no idea where I was. Forcing myself to calm down before I made any more stupid mistakes, I decided it was safest to retrace my steps. After returning to the creek and starting over, I made it back to the truck easily and called it a day. Once back in the safety of my home that night, I had a good laugh with Marshall about my misadventure.

Now that I am older and hopefully wiser, do less fieldwork, and run into fewer gators, my occasionally debilitating fear of alligators has metamorphosed back into a healthy respect. And thankfully, my alligator nightmares rarely visit me now. To my surprise, I recently learned that a group of alligators is called a congregation. When I think of the word *congregation*, I think of a gathering of people at church—peaceful, pleasant, and calm. But while my fear of gators has lessened, I don't think I will ever feel peaceful, pleasant, or calm around them—especially if we're all in the water!

Lyla McCartney

Alison McCartney fell in love with the woods and its creatures while living in rural Connecticut as a young girl and decided at fourteen that she wanted to be a wildlife biologist. She grew up in Southern California and earned a bachelor's degree in biology and a master's in environmental science. Alison has been a wildlife biologist for more than two decades, working with many different species, but has spent most of her career studying bats. She is the chair of the Mississippi Bat Working Group and a wildlife biologist for the US Fish and Wildlife Service's Mississippi Ecological Services Office.

Alison and her husband, Marshall, have three teenage girls: Maren, Lyla, and Sabra. They live in a log home on a ten-acre tract of woods in rural Mississippi with their Teddy Roosevelt terrier and Siamese cat.

Grappling with the Hands of Death—Or Not

Emily Imhoff

Growing up in a very rural part of the Appalachian region, I spent a lot of time swimming in lakes, walking in streams, and exploring natural and human-made water bodies. I just loved being around water and all the fascinating animals and plants that lived in and around it. So it seems odd that I've long been plagued by a fear of hidden underwater objects. But here I was, an aquatic biologist in my mid-twenties, and the mere thought of brushing up against a buoy chain, a submerged shopping cart, or an unseen algae-covered branch was enough to give me the heebie-jeebies.

I was working as a technician for a graduate student who was studying rare freshwater mussel species in a fairly slow-moving river near the Kansas-Missouri border. The murky water was thick with silt, and the channel was full of large, long-dead, downed trees that were mostly submerged. In the misty mornings, we'd paddle past their eerily protruding, barkless limbs on the way to Abby's research site, and I'd think about how glad I was to be up in the canoe instead of down in the watery grave of those slippery, grasping, many-branched underwater nightmares.

Luckily for me, the research site was shallow and devoid of submerged trees. Most of the autumn leaves had fallen from the hardwoods that crowded the riverbank, allowing us to bask in warm sunshine despite the cool air and cold water. We'd spend a relaxing day gathering information about habitat, measuring caged mussels to monitor their growth, hunting for additional mussels along established transect lines in the chilly water (Abby), and struggling to properly adhere numbered identification tags to the shells of live mussels (me). In the late afternoon, we'd paddle back upstream to the canoe put-in, where our ancient field truck waited to carry us to dinner.

One day, the wind was really picking up as we paddled back to the put-in, hauled our canoe onto the shore, and started unloading our gear. We made repeated trips carrying equipment up the steep embankment to the truck. I had just deposited a load and was almost back down to the canoe when a gust of wind swept it off the bank and into the river. I was stunned to see an aluminum canoe blown away like a leaf in the breeze.

Well, I'll just wade in and get it, I thought, hurrying down the riverbank. I didn't particularly want to get wet, as the water was cold and my oversized men's wet suit—stuffed with towels to fill the gaps—wasn't as effective as it could have been. But we were finished for the day, so wading in a bit would be no big deal since I could quickly warm my wet legs in the truck afterward. As I approached the shore, though, I realized how badly I'd miscalculated. The wind had continued to blow fiercely, and the canoe was rapidly making its way to the deep center of the river—and heading downstream. In fact, if it went much farther it was going to end up right over a mass of huge downed trees. An alarming collection of branches protruded from the water's surface, like the arms of patient demons waiting to subject me to all that lurked in the slimy depths.

I hesitated at the shoreline for a moment and considered my options. Abby was still up at the truck, possibly out of earshot, and by the time I went to fetch her the canoe would be far downstream. I could run along the riverbank and see if the canoe came closer to shore, but the wind direction made that unlikely. I saw only one option.

I don't remember getting in the water, just the moment I gritted my teeth and decided I had to. As I swam as hard as I could toward the canoe, the cold water filled my too-big wet suit and chilled me, though my life jacket added a little warmth while keeping me buoyant. At least I was going downstream, swimming with what little current there was. But the canoe was being pushed by the strong wind as well as the current, and we were getting farther and farther from the put-in and closer and closer to the menacing trees. Almost to the canoe, I was overjoyed that I hadn't yet felt the ghoulish arms of underwater branches touching me with their loathsome caresses.

I heard Abby yelling from the shoreline, and I tried to yell back, explaining what had happened, that I would get the canoe, and that she should *save herself* and wait on shore. God forbid anyone *else* should have to swim out here. Of course, she, like most sane aquatic biologists, probably wouldn't have minded if branches bumped against her wet suit, or if her foot touched a tree trunk lurking underwater, but that realization was beyond me. The worst part was knowing that even if something *did* brush against me, there was nothing I could do about it, because I still had to retrieve the canoe. So I was at the river's mercy, its slow but steady flow and my own traitorous limbs pushing me inexorably toward those slithery tendrils of hell.

Reaching the canoe at last, I clung to the side and scrabbled around for the end of the tow rope while trying to keep my legs as high up in the water as possible. We were still drifting along, basically among the trees now. Once I had the rope, I struggled with it for a moment before looping it over my shoulder and securing the end of it firmly in my teeth so I would have all four limbs free to swim as hard as I could to the put-in. After several moments of thrashing upstream, I looked toward the shore and realized to my horror that I'd barely made any forward progress at all.

Wow! I thought, a bit panicked. *I know I'm pulling the boat against the current, but we're hardly moving at all!*

I looked back. In a cruel twist of fate, the rope, which was tied to the bow to facilitate easy pulling, had been inadvertently looped around the canoe's center bracer bar, causing the craft to turn sideways. I was now pulling the boat broadside into the wind. *No!*

Suppressing my urgent desire to escape the sunken trees, I turned and reeled in the line until I could grasp the canoe again and unloop the rope from the bracer bar. All the while, I was drifting downstream, hoping to avoid the slippery limbs that surrounded me.

Once the rope was sorted out, I made good time swimming back toward the put-in. The water in my wet suit had finally warmed up, I'd left those dreaded trees behind, and I was feeling good. Here I was, saving the day while having evaded *every single* underwater branch. It was practically a miracle.

We continued working on the river for the rest of the fall without any more escaping-canoe incidents. I'd learned my lesson: the nightmarish dangers of the murky underwater realm always lay in wait, and I might not be so lucky to escape the hands of death next time. From then on, we tied up our wayward canoe every time we came to shore!

Run for Your Life

Elizabeth Crowe

The most frequent question I've been asked about my fieldwork as a plant community ecologist is "Aren't you afraid of bears?" I never have been—except during one field season up north while I was in graduate school.

That summer, I worked for the Tongass National Forest in Southeast Alaska. It was my first experience living so far north. By June, darkness fell at 11 p.m. and the sun peeked through my window shades at 3 a.m. The shimmering lights of the aurora borealis were a routine spectacle during those short nights. In Sitka, where I was stationed, bald eagles flocked to the shore like American robins flocked to my suburban neighborhood in the Lower 48.

Our crew hiked through wet forests and bogs, stopping at selected sites to collect data on vegetation and soils. Encased in thick wool army-surplus pants and jacket under heavy rubber rain pants and an anorak, I slipped and stumbled along in my red rubber boots, crawling over slick logs the size of small cars. Days were so cool we built fires to keep warm while we stood and ate lunch. While the other crew members characterized soils at the sites by digging forty-inch-deep pits and describing the physical features of the freshly exposed soil layers, I described the vegetation. I recorded the abundances of all plant species within a circle twenty-four yards in diameter centered on the soil pit.

The Tongass was the wildest place I had ever been. I loved that the wildlife was so conspicuous. A porcupine lumbered into my vegetation plot one day and climbed up a tree to take a nap about ten feet from me. Three beavers slapped their tails and swam in a large circle one evening as I strolled along their pond's edge. And a pod of orcas cruised by one afternoon when I was with fellow employees at a barbeque, where we ate not the usual hamburgers and hot dogs but salmon and halibut from the freezers of these full-time Alaskans.

Southeast Alaska is renowned for its brown bears—the coastal version of the grizzly bear—and we were certain to come across them during fieldwork. They are large and powerful, shockingly fast for their size, and can easily kill a person. As part of our safety training, we were required to learn how to shoot a .375 rifle, and one member of our crew had to carry a rifle in the field every day for protection in the event of a brown bear attack. Bear (pepper) spray had not yet been adopted as standard equipment.

During the safety training, I remember thinking how ineffective I would be at shooting a bear while under duress. To load and aim the rifle and fatally shoot a charging brown bear seemed an unlikely scenario, but I said nothing, did my target shooting, and begged off actually carrying the gun in the field. And despite our preparations for brown bears, I was told that bad interactions with bears usually occurred with black bears because when in brown bear country, people were less cautious about the smaller black bears.

Most of the Tongass is accessible only by boat, helicopter, or airplane, and we often flew from Sitka to project base camps in old de Havilland Beaver float planes. The planes were loud, but the water landings were surprisingly smooth. From our base camps, we flew by helicopter out to our field sites. Most of our helicopter pilots had flown in the Vietnam War, and I felt safe with them, knowing they had survived wartime flying conditions. Besides, they were forbidden from flying us in bad weather. In fact, we spent a couple of cold, tense hours on a river bar one evening waiting for the fog to lift before dark so that we could get a helicopter ride back to our bunkhouse rather than having to camp there that night.

Late that summer, during the salmon spawning season, I joined a hydrologist and his summer intern on an extended trip to survey a large river north of Juneau. We motored out of Juneau on an incredibly loud diesel-powered boat with two cabins that the three of us and the boat captain would sleep in at night. The captain anchored in a calm inlet near the mouth of the river. Over the next few days, we documented the physical features of the river valley and inventoried the associated vegetation. This glacier-fed river flowed through numerous shallow braided channels—separated by large gravel deposits—across a rocky valley bottom. The channel banks were thick with Sitka alders growing in dense, leafy thickets about fifteen feet high, which hindered our walking and drastically limited visibility to only a few feet.

When the salmon are spawning, brown and black bears are drawn to rivers, where they fish for the adult salmon that are heading upstream to spawn and die. I was told that brown bears often gorge themselves on fish for several hours in the morning and then dig out depressions near a channel bank to accommodate their swollen bellies while they lie down for an afternoon nap. As we made our way up the river each day to the section we were sampling, we walked along channel banks littered with rotting salmon, each with a single bite taken from it—evidence that we were visiting during the height of spawning season when salmon were so numerous that bears consumed only the most energy-rich portions of the fish. Nervously I thought to myself, *If the salmon are this abundant, so are the bears!*

After we took measurements of an individual channel—its width, slope, shape, and the depth changes from one bank to the other—I recorded the abundance of each plant species along transects that ran perpendicular to the channel. As I plunged through the dense stands of Sitka alder, making as much noise as I could, I looked around warily, muscles tensed, expecting with every step to come face-to-face with (and awaken!) one of the sleeping brown giants. I was always relieved when that didn't happen.

We boated to some of our sampling sites in a motorized rubber raft. To get to others, we were picked up and dropped off by helicopter. The pilot often regaled us with stories of his recent wildlife sightings, mostly of moose and brown bears. During one flight back to our boat, as the hydrologist and the pilot discussed the next day's sampling location, the pilot pointed through the cockpit window and said, "I saw an enormous brown bear along that section of river today!"

"That's exactly where we plan to sample tomorrow," replied the hydrologist, cool as a cucumber. I was anything but cool.

After we departed the helicopter the next morning, I took a picture of a brown bear paw print that was twelve inches long, which made the day all the more unsettling. During my vegetation survey in the afternoon, I came across a small tributary pool with three mating pairs of gorgeous sockeye salmon swimming over and around their gravel spawning beds. The salmon were in their breeding finery—with brilliant red bodies and green heads—and the light was just right for photographs. When we finished our work, the three of us headed back to the pool. Camera in hand, I stood in the middle of the pool about four feet from the bank, while the intern stood nearby in the shallow pool's outlet. Meanwhile, the hydrologist watched our activities from the pool's opposite bank.

I was taking pictures of the fish when the hydrologist barked, "Holy $#!t folks, don't even move!" I looked up and saw a young black bear at the end of the pool, no more than ten feet from the hydrologist. Given its small size, the bear was probably spending its first summer away from its mother. Oblivious to our presence, it was absorbed in getting a drink and looking into the water for fish. It likely had never seen, heard, or smelled a human before.

Strangely, the first thought that ran through my head was that the bear might wander over and start sniffing me like a dog would and that I would have to remain still and not panic. I have no idea where this thought came from, except that this small, young, seemingly inexperienced black bear brought to mind a Newfoundland rather than the brown bears we'd been expecting to encounter.

"Get the gun!" the hydrologist ordered. To our surprise, the bear didn't react to what was probably an unfamiliar sound. The intern dutifully walked back to where we had left our rifle and brought it to where she had been standing. Focused on the water, the bear didn't even notice her departure and return. What the intern was meant to do with the gun was a mystery to me. We all looked at each other for a few seconds. Then the hydrologist took a step along the bank, rustling the sedges and snapping a twig. The bear's head shot up. Looking wildly at the hydrologist, it turned and bolted into the forest.

That was the only face-to-face experience I had with a bear that summer. All the brown bears we saw were far enough away—bellied up to and munching on clumps of sedges or strolling along the beach turning over rocks in search of insects—that we were able to just enjoy watching them. Our rifle was merely a superfluous piece of field gear. When people ask me about working around bears, I'm happy to tell them that although they sometimes made me nervous in Southeast Alaska, in my only close encounter with one, the bear ran for its life.

Courtesy of Elizabeth Crowe

Elizabeth Crowe works as a contract ecologist doing fieldwork and data analysis on vegetation and soils for several federal agencies. She is currently analyzing historical data for use in shrubland and grassland management in southwest Montana.

Elizabeth grew up in the Midwest, where she developed a passion for the natural world in early childhood and decided in high school to pursue a career in ecology. She has a bachelor's degree in botany and zoology from Miami University and a master's in environmental science from Washington State University, and has been a Certified Senior Ecologist with the Ecological Society of America for more than two decades.

Elizabeth's hobbies include birding, hiking, traveling near and far, playing piano, gardening, baking, and soccer, including playing, coaching, and refereeing. She has lived and worked in Austria, France, Italy, Oregon, Washington, Alaska, and Montana. Elizabeth lives in Colorado with her husband and son.

A Nocturnal Jungle Encounter

Lisa Osborn Koitzsch

The pressure in my bladder was becoming intolerable. I groaned quietly and squirmed. I was going to have to crawl out of my small cocoon and brave the long, dark walk through the jungle. Sitting up abruptly, I reached for the headlamp hanging on a hook over my head. I was encased in a cotton sleeping sack inside a blue nylon tent deep in the tropical forest of southern Belize. Water murmured over small rocks and pebbles in the stream below our camp. The buzzing, burring, and whirring of the jungle's nocturnal denizens was an undulating crescendo punctuated by an occasional rustle in the leaf litter or sharp bark from a vigilant howler monkey in the canopy above.

I looked at my watch. 1:45 a.m. Three more hours before our alarms would go off and we would head out for another day of mist netting migratory birds for a Smithsonian-sponsored seasonal habitat study. I turned on my headlamp and, as its bright light illuminated my small nylon dome, stifled a startled scream. One of the many tarantulas that lived in the giant-spider-sized burrows around my tent platform had crawled up the outside of the tightly zipped door. Its enormous shadow peered menacingly down at me. Frozen, I stared up at it and briefly considered my options. Desperation won. I smacked the side of my tent, catapulting the hairy beast, and heard it hit the wooden platform. Quickly pulling on my rubber boots, I grabbed my sweatshirt and unzipped the tent. I cringed as I stepped down onto the grass, terrified that a large, hairy arachnid would emerge from the shadows and pounce on me.

"Just move. Don't look around. Shine the light on the ground. Keep moving," I muttered to myself under my breath as I passed by my sleeping field partners, all safe in their colorful little nylon worlds. I frowned as I passed Liba's tent and saw that she had left her shoes outside her tent . . . again. I fervently hoped she would check them for scorpions this time before putting them on. I was paranoid and kept everything inside my tent, including my muddy field boots.

"Keep moving. Where's the trail? Oh my god, I can't find the trail. Phew, there it is. Keep moving. Watch for snakes. Don't touch the vegetation." Each nervous step propelled me closer to the relief I was seeking, but navigating the darkness was causing my imagination to run away with me. Noises were magnified, vines reached down to grab me, roots purposely tripped me. I moved forward cautiously, intent on my destination.

I entered a small moonlit clearing surrounded by tall canopy trees and dense brush. The rickety one-sided lean-to was hidden in the shadows. A sparkling ceiling of stars drew my eyes skyward, replacing my terror with wonder. But then a loud screech brought me jumping back down to earth, and I scurried toward the wooden structure. After sweeping my light around the area, I went up the stone step leading onto the platform. Hands shaking slightly, heart thumping in trepidation, I tried to ignore the darkness and the myriad jungle sounds echoing in stereo around me. The faster I got this done, the faster I could run back to the safety of my tent.

A quick glance at the twelve-inch-diameter hole in the middle of the rough plank revealed no unwanted crawling visitors, and I turned quickly, dropped my sweatpants, and sat. *Ahhh, relief.*

Suddenly, a mad scrabbling and hissing sound erupted directly below me. I shot up off the plank and propelled myself off the platform and into the clearing. My heart pounded in my ears, and this time my scream was not stifled. Breathing rapidly, I spun around, desperately pulling up my sweatpants as I stared into the darkness. My headlamp illuminated the lean-to, and for a moment all was quiet. I was a stone statue.

Then I started to giggle. Bending over, I held my stomach and chortled. A small black face with large white spots on its forehead and cheeks peered at me indignantly. Sparkling black eyes, rounded fuzzy ears, and a pink button nose. The little creature stared at me for several seconds and then pulled its black-and-white body deliberately up out of the latrine and shook its head. It stalked across the wooden plank, dropped down onto the platform, and waddled toward me.

At that moment, I realized what I had just done. Turning quickly, I flew back down the narrow trail, headlamp bouncing wildly, to avoid receiving a return gift from the southern spotted skunk I had unintentionally peed on.

Hope and Heartbreak

The Last Wild Condor

Janet A. Hamber

Few of us are present at the end, when an iconic species is on the brink of extinction and the actions we take might make all the difference. By the 1980s, the California condor—a majestic, soaring scavenger with a nine-foot wingspan—had faced catastrophic losses. In 1982, only twenty-two California condors remained in the world. To those of us who fought to save the species in those desperate years, the pressure was intense, and the journey was an emotional roller coaster.

My involvement with condors began in early January 1976, when I accompanied a biologist who wanted to observe a possible pair of nesting condors in Southern California's San Rafael Wilderness, part of the Los Padres National Forest. Moments after we had settled into our observation site, I yelled "Condor!" and watched a pair swing around from the west and then float below us before landing on a rocky outcrop. They were magnificent—and I was hooked. For the next decade, I monitored the nesting activities of this pair, which became known as the Santa Barbara pair, and helped with research activities initiated by the Condor Research Center, founded in 1980. A growing number of us were monitoring the few condors that remained in the wild, while also trying to determine why the birds were disappearing.

I first saw the bird that would become famous as AC-9 (for Adult Condor Number 9) in 1980, as an egg tucked into a cave on Redrock Mountain in the Angeles National Forest. New colleagues had found the nest site in March, after spending several days watching a pair of condors in the Redrock Canyon area. I remember being astonished, and a little embarrassed, to learn that the biologists had found the nest cave so quickly, on one of their first forays into the field. I wondered if they would think I was an incompetent observer since I typically spent weeks each year searching for the Santa Barbara pair's latest nest cave and occasionally failed to find it at all. But I tried to reassure myself by considering how remote and inaccessible much of the Santa Barbara pair's nesting territory was.

On March 30, 1980, I joined biologist Noel Snyder on a trip to the Redrock nest area. As I later wrote in my field journal, "What an amazing day." I got up at 3:35 a.m., left home at 4:28, and met Noel in Ojai, a small city in Ventura County, north of Los Angeles. We hiked up a road that ran along Castaic Creek to the designated observation point at Redrock Mountain, arriving at 8:50. At 9:33 a.m., we spotted an adult condor flying in from the west. After making several short flights in the vicinity, it landed at the cliff nest site. As it did so, a second condor came out of the cave and flew off, while the returning bird entered the cave at 10:05. The nest exchange we had just witnessed meant the birds were incubating an egg.

We left our observation point and moved up the road and off into the chaparral, hoping to get a better view into the nest cave to see the incubating condor. To our delight, from our new vantage point the bird was almost in full view, a rarity in my experience. At 5:17 p.m., it began to pick up sand and pebbles, dropping them around itself to build the rim of its "nest," a slight depression scratched into the floor of the cave. At last, when the condor stood up, I saw a pale greenish-blue egg in front of the condor's legs. I had spent many long days watching the Santa Barbara pair's distant nest but had never been treated to the unobstructed sight of a condor nest exchange, an incubating adult, or an egg. What a thrill!

We left the area, exhausted, at 5:30 p.m. It had been a long but unforgettable day, and as I noted in my diary, "Fantastic!!"

Later that summer, from July 25 through 27, I again monitored the Redrock nest. The nest site had two entrances that were connected by a rock tunnel, into which the condors sometimes disappeared for hours at a time. By now a blind had been set up that consisted of used bed sheets—gathered from a thrift store—draped across tall yucca stalks. (Whoever said the government spends money lavishly?) It was equipped with two chairs and a small table.

The sun rose over Redrock Mountain at about 7 a.m. and beat down on the blind all day, sending the temperature soaring above 100 degrees. But the excitement of being able to watch and gather data as the condors raised their nestling kept nest watchers riveted for twelve-hour stretches despite the uncomfortable circumstances. Moreover, we always had a chance of learning something new that might help to save the species. (Indeed, some of what we learned proved critical to restoring condors: that the birds would lay a replacement egg if their lone egg was destroyed, that a pair would lay in successive years—rather than once every other year—if their nesting attempt failed, and that raven predation could threaten condor eggs.)

My next opportunity to view the Redrock nest came in late September. The Redrock chick was now moving around the nest site—walking along cliff ledges and clambering onto nearby rocks—but not yet flying. By the time I left the area in early November, the full-grown chick had spread its giant wings and taken its first short flights in the vicinity of the nest.

By 1982, when only twenty-one condors remained in the wild (and one in captivity), the condor known as the Redrock chick had become a black-headed juvenile that had been trapped and outfitted with a radio transmitter. In addition to monitoring the wild birds to try to determine their fates, the condor recovery program had initiated a captive-breeding effort in the hopes of bolstering the wild population with future condor reintroductions.

Given that captive breeding was still a new conservation tool, these measures divided the conservation community. Some raged that a caged condor was not a condor at all but an "uninspiring, pitiful, ugly" creature. They felt that "the beauty of a California condor is in the magnificence of its soaring flight" and believed that capturing wild condors for a captive-breeding program would ensure the species' extinction. Others believed this last-ditch effort was essential to reversing the condor's declining numbers. Meanwhile, with the success of the captive-breeding program still far from certain, wild condors kept dying or disappearing.

By the end of 1984, only fifteen wild condors remained, but we biologists were hopeful because we had documented five breeding pairs. Our mood was one of cautious optimism as we gathered to prepare for the yearly search for condor nests at the beginning of the 1985 nesting season. A successful season would give the condor population a much-needed boost and likely forestall the species' extinction. Biologists would take each pair's first egg and raise it in captivity, and then we would monitor the outcome of the replacement egg, which would be raised in the wild. With pairs "double-clutching," or laying a second egg to replace the first, we hoped to double the condors' reproductive output. If we were successful, several of the offspring from eggs that were taken into captivity would be released in 1986 to augment the wild flock.

But that hopeful plan was not to be. As nest watchers fanned out to search for nest caves in late January 1985, reports filtered back that either one or both members of pairs were missing from their breeding territories. By April it was clear that disaster had struck. Six condors were missing. Only nine birds had survived the winter. Worse yet, only one intact breeding pair remained: the Santa Barbara pair. Our plans were shattered. Anxious about the safety of the remaining condors, the US Fish and Wildlife Service authorized the trapping of three more condors for the captive-breeding program. Now, only six birds remained in the wild—two females and four males.

In January 1986, AC-3—one of the two females and a member of the last breeding pair—died of lead poisoning. More rancor and dissension broke out among biologists and conservationists. Some wanted to capture the remaining condors to keep them safe and bolster the captive population. Others thought some condors should remain in the wild to justify protecting their habitat from development and to serve as mentors to any captive-reared condors that might be released to the wild in the future.

During the early months of 1986, the three surviving adult males and the Redrock youngster, which had just reached adulthood and been dubbed AC-9, all pursued AC-8, the lone female left in the wild. Spurning the older, more experienced males, she eventually chose the young AC-9 as her new mate.

Tragically, the shell of their first egg was so thin that it was crushed. AC-8 laid a replacement egg, which survived but was taken into captivity and later hatched at the San Diego Zoo. Given the condor's now-perilous status, the decision was finally made to capture the remaining five condors and bring them into captivity. AC-6 and AC-8 were trapped that summer, and AC-2—the Santa Barbara male that I had watched at his various nest caves for eleven years—was trapped in December. Now only two condors remained in the wild.

In February 1987, male condor AC-5 was captured with a cannon net—a large, folded net attached to projectiles inserted into small "cannons." When the cannons were fired, the projectiles flew out over the target, dragging the attached net. For trapper Pete Bloom, it was a poignant moment. As he placed AC-5 in a transport kennel, he noticed AC-9, the very last wild California condor in the world, watching him from a perch in a large oak tree above the trap site, his body silhouetted against the setting sun.

That night we set out carcasses at several potential trap sites, as we had each night for the past year. The sites that we baited were determined by what the "chase biologist" of the day had found. Armed with radio telemetry equipment, the chase biologist began each day where a condor had roosted the night before. When morning came, the chaser followed the bird by truck and foot throughout the day. If the condor ended up at a potential trap site, we'd alert the trapping team, which would set up a capture net under the cover of darkness to trap the bird the following day. The chase for AC-9 was on, and the pressure was intense to trap him before he disappeared or died in the wild. But for months, AC-9 ranged widely and eluded us.

I began my turn as chase biologist on April 18 at the Hudson Ranch, now Bitter Creek National Wildlife Refuge, where AC-9 had been seen the previous day. I watched him approach one of the bait carcasses and spend three hours near it, but he was chased off by golden eagles and left without feeding. Still, he was likely to return to it the next day. So now I had to do my job—if I could overcome my reluctance to condemn AC-9 to captivity. I considered not calling Pete Bloom. Even if I did call him, I reasoned, he was unlikely to be able to organize a trapping team at this late hour to meet at 4:30 the next morning—on Easter Sunday. Maybe AC-9 could remain in the wild just a bit longer. But I couldn't neglect my duty. With a heavy heart and much misgiving, I made the call.

In the dark that early Easter morning, the trapping team assembled at the ranch's barn. I remained on the roads near Hudson Ranch to report AC-9's movements. At 8:47 a.m., I picked up his radio-transmitter signal coming from nearby Brush Mountain and listened as the signal moved westward toward Hudson Ranch. By 9:48, he was at the trap site. And edging closer to the carcass.

Get out of there! Get out of there! I screamed to him silently, willing him to get out of harm's way. Scarcely aware of what I was doing, I jumped out of my truck, banged the door, and waved my hands, knowing I was too far away for him to see me, knowing I was powerless to help him. I knew I was being unprofessional, but I couldn't suppress a decade's worth of instincts to do whatever I could to protect condors from harm. Climbing back into my truck, I watched through my spotting scope as AC-9 approached the carcass, grabbed a quick bite, retreated, then sidled in for another bite.

Then I saw the puffs of smoke from the cannons, and the net shot over AC-9, trapping him as he scrambled to escape. My heart sank into my gut like a rock, my head dropped down to my chest, and a wave of sorrow swept over me for all the lost hopes and failed efforts, all the bitter infighting among conservationists, all the callousness of those whose actions had harmed condors and their habitat. Above all, though, I mourned the loss of the great wild freedom the condor had represented. And I mourned the end of my condor career. My opportunity to experience the wild world, so much of my sense of being *me*, all of it gone in a puff of smoke. I put my head down on the steering wheel and cried.

A short while later, I drove down to the trap site, where I saw AC-9 standing behind bars in his kennel, awaiting transport to the Los Angeles Zoo. I told him I was sorry. Pete Bloom was equally sober. We made a quiet promise to AC-9 that if the zoos didn't one day release condors, we would scale the fences, kidnap him, and release him back to the wild.

Eleven years of struggle, work, and hope to save wild condors were over. No amount of habitat protection, laws, education, or diligent observation had stopped the bird's slide toward extinction. We had started too late, and time had simply run out. As I drove home, I passed area after area where I had watched condors—places where the species had persisted for tens of thousands of years: Snail Canyon, Madulce, Sespe Creek, Pine Mountain, Cuyama Peak. All seemingly empty, desolate, and devoid of life now that the great wings of the California condor no longer graced the skies.

Adapted from my diary:

> January 4, 1988. I've had it on my mind to write up the last days of wild condors in California for months, but somehow I couldn't do it. It is still painful for me to think about those huge wings held captive in a cage in LA's smog and know that never again will I thrill to the sight of that majestic bird soaring for miles so effortlessly on the wind currents. I had said I didn't want to see the last few wild condors trapped, but maybe it was just as well I was there. Rather like going to a funeral and seeing a loved one interred—the absolute finality of it all is overwhelming . . . The end of an era . . .
>
> [As AC-9 headed for the zoo] I just knew that I had been part of a historic moment—the purposeful removal of an endangered species from the wild to save it; but can the habitat be saved without the species, and will they ever release condors? The zoos insist they will . . . I can only hope captive breeding will work and the resolve to return condors to the skies of California will be maintained. FINIS.

In 1992, five years after AC-9 was trapped, the condor recovery program began releasing captive-reared condors to the wild. Gradually, over the years, the free-flying population grew despite the perils that awaited them in the wild. Biologists eventually discovered the smoking gun that explained the species' decades-long demise: condors had been dying of lead poisoning after ingesting lead-bullet fragments in hunter-killed carcasses. Despite that ongoing threat, pressure mounted to return the last wild condors to the skies from which they had been taken.

Adapted from my diary:

> May 1, 2002. AC-9, twenty-two years old, was released back to the wild today, after fifteen years in captivity. I brought my son, Bob,

his wife, and my ten-year-old grandson, Zachary, along to view the release. We joined some of the old-timers, like Pete Bloom, from the 1980s. Over one hundred people gathered to view the release. The net covering the release pen in the Sespe Condor Sanctuary was opened at a corner . . . and we waited, and waited. After two hours, an intern finally gave AC-9 a push toward the opening. Anybody who expects the condors to rush out to freedom is due for a disappointment. But according to newspaper articles, upon seeing him fly again, I exclaimed, "He's free. Free, free, free, free at last! It's incredible that he's back out. It's a wonderful feeling."

From my field journal: "Since his release, AC-9 has adjusted well to life back in the wild. There has not been much interaction between him and his wild mate from 1986, AC-8." (AC-8 was released back to the wild in 2000 and had not chosen a new mate. She was shot by a man hunting feral pigs with his dad in February 2003.)

I continued to go out to the release site at the Hopper National Wildlife Refuge to monitor all the released condors three or four days a month in 2002 and 2003, often seeing AC-9 or hearing the blips from his radio transmitter.

In early 2004, AC-9 paired with Condor 192—a captive-reared condor released to the wild at Big Sur, in central California—and began nesting in a remote portion of Southern California's Agua Blanca drainage. Regrettably, I had to give up watching him. Now in my seventies, I couldn't make the arduous journey to his far-away nest site.

But on Easter Sunday, April 11, 2004, seventeen years after AC-9 was taken into captivity, I celebrated the hatching of his chick, which emerged from its egg in the wilderness to which his father had so recently been returned.

In the ensuing years, AC-9 nested again and again. When his mate died in 2015, he paired with another captive-raised female that had also been released to the wild near Big Sur.

On June 29, 2016, AC-9's transmitter signal remained stationary. Several days passed before biologists could hike in to find him. By the time they arrived at the location from which AC-9's last signal had transmitted, there was no trace of him. No transmitter. Not even a feather to mark his gravesite. AC-9 was thirty-six years old. He was known the world over. The last wild condor. The last of an era. And a regal symbol of new beginnings and second chances.

Courtesy of Santa Barbara Museum of Natural History

Janet (Jan) Armstrong Hamber grew up in Douglaston, New York, where her father instilled in her an interest in the natural world during walks through the neighborhood. She entered Cornell University in 1947 as one of about 40 women among the 250 men admitted to the College of Agriculture. She earned a bachelor's degree in botany, though taking Dr. Arthur Allen's ornithology course in her final semester sealed her interest in birds. Jan married and traveled all over the US for a couple of years with her husband, living in Connecticut and then California. The day her son entered kindergarten, she went to the Santa Barbara Museum of Natural History to volunteer in the Department of Vertebrate Zoology.

September 2024 marked Jan's sixty-fifth year with the museum, where she served ten years as a volunteer and seventeen as a curator, and is now the California condor biologist and archivist. She is establishing a fund to preserve the archives she created to document the condor's near extinction, the race to save the species, and the successful recovery program so that future students will understand the effort that went into preserving this critically endangered species.

Alpha Female 1340

Julia B. Smith

Using my hand to shield my eyes from the blinding morning sun, I paced the tarmac. My breath steamed in the frosty winter air typical of a January dawn in Arizona's White Mountains. It was a perfect day for flying. Bright sun, clear blue sky, cold enough that the snow on the ground wouldn't melt, but no hint of a fresh storm that might sock in the airport and ground us. My tension built as the minutes before I had to buckle myself into the front seat of the helicopter slipped away. Even though I was relatively seasoned by then, I never flew without a healthy awareness of the dangers our helicopter work entailed.

The best I could do was prepare. Prepare by scrupulously following the checklists—of gear, tasks, and individual animals targeted for capture—that I had developed for our team. Prepare my body for the physical exertion that lay ahead. And prepare my mind, steeling it against fear and doubt. Now, only moments before the day's operations were to commence, I had nothing left to do but trust myself, tapping into my tenuous confidence the way one might gingerly nurture a weak flame that could disappear in the slightest breeze.

"You look like Rambo!" shouted the helicopter pilot, disrupting my thoughts. I grinned in response to the thirty-something-year-old man I had been flying with for years as part of the Mexican Wolf Recovery Team. I was wearing a green headband to control my long dark hair in the wild winds that would blow in my face as we flew with the doors off. My boss, who would be darting wolves from the back seat of the helicopter that day, came over to join in the playful ribbing over my current state of dress: heavy black leather ankle-high fire boots laced all the way up; army-green flight suit a little on the baggy side, its pockets stuffed with wolf muzzles and hobbles; bright-orange vest so I could be spotted against a backdrop of snow or thick timber; white helmet under my arm; and of course, the green headband. Despite an agency culture that typically resulted in my male colleagues playing this role, I had infiltrated the macho cowboy clique of tall, strapping men who pursued wild animals at top speed from the air, one of the most dangerous (and glorious, to some) tasks a wildlife biologist could undertake.

When you are capturing wildlife from a helicopter, there are Good Days and Bad Days. Bad Days are the ones when you are grounded early because of weather or mechanical failure, or the animals are in terrain so rugged or timber so thick that attempting to dart one of them would be pointless. Of course, in a helicopter, days can get a lot worse than that—airplane and helicopter crashes are the leading cause of job-related deaths for wildlife biologists.

But then there are the Good Days. Cloudless, clear blue skies. No wind. Helicopter in flawless condition, honed as a chiseled wildlife-capture tool, unparalleled in its precision with a good team on board. Targeted animals out in a flat plain or alpine meadow where their speed on the ground is no match for the machine in the air. When the three of us spooled up into the sky and pursued our first targets of the day, the Diamond pack wolves, I knew it was going to be a Good Day.

These winter helicopter captures presented one of our best opportunities to gather information that we hoped would support efforts to inch endangered Mexican gray wolves a little closer to recovery. The population of this rarest and most genetically distinct subspecies of gray wolf in North America was hovering at just around one hundred animals at the time. Obtaining as much information as possible about each pack was important to understanding how well the population was recovering.

I directed the pilot toward our quarry, navigating by the beeps—signals from a radio-collared wolf in the Diamond pack—that were sounding louder and louder as we flew closer and closer to the animals. I knew the country that stretched out below me better than I knew my hometown. Years of following wolves had taken me on every back road and trail, from creek bottoms to butte summits. I knew the Diamond wolves would be in a great spot for capture—the wide-open pinyon-juniper flats.

I adjusted my radio telemetry receiver as the beeps in my ears grew louder, and then thunderous. The wolves had to be close.

"We're over them now," I told the pilot and my boss, who handled the dart rifle. "Start looking."

The large pack of wolves was fairly easy to pick out once it started moving away from the helicopter's deafening rotor noise. As the pack made its way between junipers, we rushed to count each individual, and I scanned them for our primary capture target—the breeding male whose radio collar no longer functioned. Identifying individual gray wolves from dozens of feet above them while in hot pursuit can be nearly impossible, but this time, I picked him out immediately. He was easily twice the size of the other wolves in the pack, and I could just make out the dark line of a collar around his neck.

"That's 1249 right there," I said, pointing out the wolf and calling him by his identification number.

The pilot deftly paralleled the wolf, keeping the helicopter just ahead of the animal's path. My boss, a practiced and talented gunner, had a dart in the wolf in minutes. Alpha Male 1249 had learned to avoid steel traps and remote cameras, and was nearly impossible to glimpse from the ground. Now he faltered and then fell as his back legs gave way to the immobilization drugs in the dart. Soon he was lying on his side, and the pilot landed the helicopter a few yards from him.

I unplugged my helmet from the communications system that passengers use to talk during flight, briefly patted my pockets to make sure my equipment would be secure in the strong wind created by the helicopter's rotors, and carefully exited the cockpit, crouching to avoid the blades still spinning a few feet above my head. I approached 1249 in near disbelief, disarmed by the presence of this powerful, intelligent pack leader who had so nimbly evaded detection and capture by other methods and was now lying helpless before me. But I had a job to do.

I touched him lightly to make sure the drug had taken effect. He was unconscious. I knelt next to him, lifted his head, and slipped a muzzle over his powerful jaws, clicking the buckle behind his ears. Next, I hobbled his legs, tying secure knots around them as I had practiced countless times, so that we could transport him. I put my hand on his chest to check his breathing and slid a thermometer under his tail to make sure his temperature was within normal range—wolves pursued by helicopter are prone to dangerous overheating. My ground duties finished, I hoisted his limp eighty-pound body into my arms and carried him to the helicopter, where my boss helped pull him into the back seat. We were thrilled that within twenty minutes of starting our operation, we had our day's first capture, especially one as significant as this breeding male wolf.

Rather than processing 1249 where he'd gone down, we returned to the sky and flew him to a site ten minutes away, where we dropped him off with one of the ground crews. The crew, which included a veterinarian, was ready and waiting to process the wolf: weigh him, replace his collar, draw blood, administer fluids and vaccines, and take measurements and photos. The crew would then drive him back to the vicinity of his capture location and release him once he woke up. This arrangement provided excellent care for the wolf while also maximizing our limited and costly flight time and thus our opportunities to capture other wolves.

After dropping 1249 off to the ground crew, we lifted off once more to return to the Diamond pack. Because the pack was in such open terrain, we knew we would be able to capture another wolf from the pack easily. "Make hay while the sun shines," my boss liked to say. Sure enough, upon reaching the pack's earlier location, we darted an uncollared yearling wolf and transported him to the processing crew in short order. "Two wolves in under thirty minutes . . . that's gotta be a new record," our pilot drawled. My boss and I exchanged grins. Not even nine in the morning and it was already a Good Day.

"We're heading to Marble next," I said.

As we crested a snow-covered ridge and I started to hear faint beeps from my telemetry receiver, we discussed our next targets—the Marble pack wolves. I had been following these wolves on the ground for the better part of a year, initially trying to determine whether a dispersing female wolf known as 1340—born three years earlier and radio-collared at nine months old—was traveling on her own or had found a companion. And if she had found a mate, did they have pups—and had those pups survived? Pups would mean a new wolf pack, and pups surviving into the new year would mean another successfully reproducing wolf family. I had been keen to find out what 1340 had to tell us.

As we flew over miles of forest and mountain in minutes, I reflected on everything I had discovered over the course of the year. The collar 1340 wore allowed me to track her—using radio telemetry—from the high peaks to the deep draws that make the White Mountains so rugged and wild. I located her in the same canyon every day that spring. I placed trail cameras in the area, and before long, the photos revealed that she was paired up with an unknown male companion wearing a nonfunctional radio collar. Capturing her on camera regularly with the same male and finding her in the same spot every day almost certainly meant she had dug a den and given birth to pups in April or May. When an animal that can easily cover fifty miles in a day slows down in spring, it means something.

My suspicions were confirmed that fall when I discovered among the thousands of photos from the trail cameras a picture of five pups following 1340 and her mystery mate. In August and September, I'd trapped and collared two of those pups, fitting them with GPS collars that we hoped would work for years, allowing us to track the surviving young wolves as they grew and dispersed.

Now instead of being a lone disperser, 1340 was the matriarch of a family. She was a breeding female member of an endangered species, a demographic any biologist recognizes as critically important to the survival of that species. The only thing we did not know about the Marble wolves was who the breeding male was—the male I had so often seen in trail camera photos, his beat-up, nonfunctioning collar teasing me with the question of his identity. And with the prospect of finally answering that question filling me with nervous excitement, I directed the pilot toward 1340, whose radio-collar signal was pulsing in my ears. I knew the male would be with her.

The beeps from 1340's collar led us toward the small community of Greer, Arizona. Just north of the tiny town, we flew over a wide, open meadow with a reservoir, a place favored by the local elk herd. My gaze immediately landed on a large gathering of ravens at the water's edge. Before even seeing it, I was certain we had come upon a fresh kill made by wolves and that we would find the Marble pack right here, in this clearing—flat, nearly treeless, and with no power lines to avoid, just about the best capture location we could hope for. It appeared our early morning success wasn't yet finished.

As the helicopter thundered toward the ravens, they rose from the ground like a black mushroom cloud and scattered to the four winds, revealing a freshly killed elk and patches of blood-stained snow. We no longer needed to navigate by telemetry. In front of us were the five wolves of the Marble pack—the breeding pair and three remaining pups, down from the five documented earlier that year—bedded in the snow near their kill.

The wolves instantly took off, loping through the snow, and scattered in all directions, as the ravens had. We could easily count the wolves, but isolating our quarry was trickier. Four of the five were collared—1340, the two pups I had captured a few months earlier, and the breeding male of unknown identity. To isolate our target wolf, we had to differentiate among the animals from the air. The breeding male was our target, and the only wolf we needed to capture from that pack. But we would have to separate the male from his family members to ensure we were pursuing the right wolf, and identifying near-identical wolves by their weathered collars and the slight differences in their size from thirty feet in the air is a crapshoot at the best of times. All we could do was give it a shot.

Often when a wolf family is pursued from the air, the breeding male and female will stick together even when other pack members break away from one another. Our pilot used this knowledge to our advantage to lock on to the two slightly larger wolves that would not split away from each other as we pursued them. This was our breeding pair, and one of them was the male we had long been trying to identify. Quickly exhausted by the chase in deep snow, the wolf pair slowed down and then stopped after only a few minutes. They stood next to each other, eyes fixed on the deafening behemoth hovering in front of them. This was the best chance I was ever going to have to pick out the male.

But even after studying all those trail camera photos of the pair, I could not do it. The wolves themselves were of indistinguishable size and pelage. The collars didn't help me differentiate the wolves, either. The beeps of 1340's collar thundered in my ears, but with the wolves so close to one another, it was impossible for me to discern which wolf was wearing the collar that was emitting the booming signals and which was wearing the nonfunctioning collar. And both collars looked identical—the same make and model, the same weathered brown, the uniquely colored and patterned duct tape they once sported long since worn away. I gulped.

"I don't know," I stammered. "I can't tell them apart."

"No problem," my boss replied. "I'll just go for one of them."

The wolves remained side by side, still facing down the helicopter. My boss darted one of them within seconds, starting the chase anew when the wolf felt the dart hit and took off, its mate following. We pursued the animals to keep them within our sight, and before long, one of the wolves began to stumble as its back legs gave out. We backed off and hovered high overhead, our eyes fixed on the wolf pair, as we watched the darted wolf steadily losing consciousness in the snow. The wolf was down.

The helicopter descended, moving in to land so I could retrieve the darted wolf, now lying on its side in the snow. The undarted wolf had not left its mate's side and remained steadfastly in place, even as the helicopter came closer and closer. As the skids touched the ground next to the unconscious wolf, the mate suddenly lunged toward the helicopter, teeth bared.

"Whoa!" shouted our pilot. "What the hell was that?"

The wolf rushed us again, snapping its jaws, defending its downed mate the best way it knew how. None of us, in all our combined years pursuing wolves, had ever witnessed a wolf trying to attack a helicopter.

"That must be love!" we heard our spotter's voice crackle over the radio. She had witnessed the entire show from hundreds of feet above in the spotter plane circling overhead.

It became clear that this wolf was not going to leave its mate's side, regardless of the rotors roaring just feet away. My boss loaded another dart.

"We're going to dart this one, too," he said.

Darting wolves from a moving helicopter is extremely difficult on a good day, and here we were with more opportunity than we knew what to do with. The wolf who had confronted the helicopter minutes before was now taking off with a dart in its hindquarter. Using my GPS, I recorded the location of the downed wolf—it likely would be unconscious for at least forty minutes or so with a full dose of Telazol—and we took off in pursuit of the running wolf, not wanting to lose it from our sight.

My adrenaline was pumping, and apparently so was that of the wolf we had just darted. Although the wolf had slowed down and begun to lose the use of its hind legs, it continued to push onward through the snow. We backed the helicopter off, rising high in the air to give the animal space and allow time for the drug to hit its bloodstream. The wolf stopped moving, but even after ten minutes passed, it still had not put its head down.

Our pilot brought the helicopter in to land, the skids sinking slightly into the snow. My boss and I carefully exited the helicopter, rotors spinning overhead, he with dart rifle in hand and I with muzzle, hobbles, and catch pole. We slowly approached the wolf, and at first, it seemed as if I might be able to muzzle and hobble it for transport. But as I inched closer, the wolf sprang up and bolted away. Instinctively, I tried to chase it down but, of course, nobody has a chance of catching a wolf on foot, even if the wolf has partially lost the use of its back legs.

When the wolf slowed again, my boss raised the dart rifle and took aim. He hit the wolf easily once more, and this time, with a double dose of drug now coursing through its bloodstream, the wolf was down within minutes. I rushed over, muzzled and hobbled the animal lying in the snow, and pulled a thermometer out of my pocket to check the wolf's temperature. As I slid the thermometer under its tail, I saw that this was the male. The mystery wolf I had been after all year. Soon, the mysteries of this pack would be solved. And that meant the other wolf we had left behind, lying on her side in the snow, was 1340, the breeding female of the Marble pack.

I squatted and hauled the wolf into my arms, staggering under his limp weight. As I waddled back to the awaiting helicopter, rotors still turning, I shouted to my boss, "It's the male!"

We both climbed into the back seat with the wolf across our laps, and before we could buckle ourselves in, the helicopter was in the air, barreling back to our previous location so I could retrieve the female. Her body in the snow came into view quickly, and minutes later I was on the ground, muzzling and hobbling her, checking her breathing and taking her temperature. I worried that her body temperature might have dropped dangerously because we had left her lying in the snow while we pursued the male, but both her temperature and respiration were within normal range. I breathed a sigh of relief as I carried her to the helicopter and lifted her into the back seat.

As we took off and sped toward the airport where the ground crews awaited our precious cargo, my boss and I shook our heads in disbelief and high-fived, not believing our luck. We proudly announced over the radio to the ground crew that we had captured not just one wolf but two, and not only that, but we had both the breeding male and female, together. Even though we hadn't initially planned to capture the female, we would capitalize on this opportunity to outfit both wolves with brand-new top-of-the-line GPS collars that would help us learn all about this pack's movements and make it much easier to determine whether the pack denned again in the future. All three of us in the helicopter grinned from ear to ear, thrilled and triumphant, our hearts pounding with adrenaline and joy. We all knew there were Good Days in the helicopter, but this was now officially the Best Day Ever.

I looked at the two large wolves sprawled across us, the wind whipping their long, thick winter coats. As I held 1340 on my lap, I pressed my hand to her chest to count her breaths yet again. My blood went cold and my heart jumped into my throat. I moved my hand around her chest, feeling all over her ribs, but there was no rising and falling, no steady intake of breath. 1340 was not breathing.

"She's not breathing," I said to my boss. His eyes locked on mine in silent alarm as his hand instantly moved to her chest. We called in to our ground crew over the helicopter's radio, "The female isn't breathing." We were only minutes from landing.

The second the skids touched the tarmac, we handed the two wolves off to the ground crews. Two veterinarians immediately administered CPR to 1340 and injected her with what would hopefully be life-saving steroids. I hung back from the crew surrounding the wolves as our pilot shut down the helicopter, its rotors slowing with a high whine. I raised my visor and removed my helmet. From a distance, I tried to discern what was happening.

Staff and interns were quickly and methodically processing the male, drawing his blood, administering vaccines, pulling off the dirty old brown collar and replacing it with a shiny new one. My eyes shifted to 1340 and the veterinarians, who were doing everything they could to start her breathing again. When the crowd finally backed away from the wolf and I could see 1340, motionless on the ground, I knew she was dead. That she had, in fact, died in my arms in the back of the helicopter, taking her final breath sometime during the mere minutes between retrieving her from the snow and handing her over to the ground crew.

I could feel hot tears welling in my eyes as my brain fogged. Whatever successes we had had that day no longer mattered, and would never matter to me again. Finally learning the identity of the Marble breeding male had lost all importance and meaning. The Best Day Ever had turned into the worst kind of day that a wildlife biologist can have—a day when we kill an animal that we have dedicated our lives to saving. Many wildlife biologists will tell you, some of them quite clinically, that occasionally losing an animal is part of the job. It goes with the territory. It happens to everyone once in a while, and to every species we work with. Although capture fatalities are exceedingly rare, I knew intimately that they were always a possibility any time we laid our hands on a wild animal.

Some wildlife biologists will also tell you that individuals do not matter. We are trained to care about populations, hardwired to monitor trends. The fate of one individual is not necessarily significant—if the overall population is trending upward, the loss of a single animal will be statistically meaningless. But my well-trained biologist's mind could not protect me from what my heart knew.

Only twenty short minutes ago, a wild wolf family was resting, undisturbed, near an elk carcass in a quiet, snow-covered valley in a national forest. Fifteen minutes ago, one of those wolves was risking his life to defend his mate from an unknown menace that must have been incomprehensible to him. Now their matriarch was dead, and no matter what I knew about the resilience of wolf populations, trends, and recovery, the loss of this individual did matter. It mattered to me. And it would matter to the rest of her pack waiting for her return.

As a young girl, I had always dreamed of working hands-on with wildlife to aid in their recovery. But that hopeful dreamer had never envisioned a moment like this one, where the creature I had followed for a year—one of a species I had fought so hard to save—took her dying breath in my arms. There can be so much power, glory, and identity wrapped up in what we do as wildlife biologists, especially when it comes to capture work that requires a high level of precision and expertise. At the time, I was the only woman in my agency performing such work, and it put me on an equal playing field with my male colleagues, if only in my own mind. To this day, I am proud of my skill set and thrilled by the experiences I've had, which are still rare among biologists—especially women. But as much as I loved what I did and was empowered by it, the eventual destruction of the Marble pack, which we unintentionally initiated that day, made me question what I was doing and why I was doing it.

The death of 1340 was a freak accident. A necropsy later revealed that the blow from the tranquilizer dart impacted one of her vital organs. If the dart had hit an inch to either side, her capture might have had a completely different outcome. We had been much closer to her than we usually were when darting wolves and hadn't made the necessary adjustments to accommodate that. Although we prepared meticulously and paid attention to the tiniest detail, unforeseen accidents can happen anytime anyone handles a wild animal. Nevertheless, I was tormented by questions. Was any of the information we had hoped to gain from deploying another wolf collar worth what we did to this female, to her mate, to her pups? Could we have gained the same information some other way, without the stress and potential harm of capture? Was whatever still remained for us to know about these wolves really worth the risks we imposed on them by interfering with their lives?

In the weeks and years following that day in the helicopter, I would capture and collar dozens more wolves, because it was my job to do so. There were a lot more Good Days, and plenty of Bad Days, too. The work was never boring and never failed to make me pause and say, "Wow." But the reasons for doing the work and the purposes it served started to unravel in my mind. Wolf biologists are constantly assailed by opposing forces, by those who worship wolves and those who demonize them, by those who insist on their protection and those who demand their extermination. Our work, too often, is buffeted by political winds that reflect these tensions.

I knew there were plenty of less intrusive ways to study wolves and other wildlife, with new methods being developed all the time. However, the information about individual wolf movements and population numbers obtained using these nonintrusive methods was less specific and immediate than that provided by radio collars and rarely satisfied either those who idolized wolves or those who feared and hated them. Nevertheless, I found myself asking if we were putting collars on wolves to answer questions that would assure their survival, assist in their recovery, and help weave back together intact ecosystems—or inserting ourselves into their stories primarily to serve human needs and desires. After 1340 died, I couldn't stop asking myself: Is it really worth it to put yet another collar on yet another wolf?

Kaden Coty

From childhood, Julia Smith knew she would dedicate her life to creating a better world for wildlife. She holds a bachelor's degree in zoology from Michigan State University and a master's in zoology (wildlife ecology) from Southern Illinois University Carbondale. She studied the recolonization of black bears, cougars, and wolves in the Midwest and then worked on Mexican wolf recovery in Arizona and New Mexico, ocelot recovery in Texas, and wild turkey survival in Missouri. Subsequently, Julia worked on wolf recovery for more than a decade, first as Arizona's Mexican Wolf Project field supervisor and then as Washington's wolf policy lead.

She is the manager of the endangered species recovery section at the Washington Department of Fish and Wildlife, where she facilitates the recovery of nearly fifty at-risk wildlife species, from butterflies and frogs to wolverines and grizzly bears. She is passionate about turning lawns into native plant and food gardens, ensuring funding for wildlife conservation, and creating space for wildlife in an increasingly developed world.

Lights on the Beach

Elizabeth Torrey

Parque Nacional Tortuguero, Costa Rica, 2011

So dark . . . so tired . . . and so many bugs . . .

Just before midnight and miles from the nearest power grid, the moon and stars shone brightly in the tropical night sky, providing the only light by which to see. All was quiet in the research camp on Costa Rica's northeastern coast, until my iPod alarm blared through my earbuds. It was time to search for sea turtles!

The thought of seeing a female green sea turtle fighting her way up the beach to lay her eggs by the light of the moon was enough to motivate me, a young biologist, to swing my legs over the side of my wooden cot and clamber out from under the mosquito netting. After scanning the floor for scurrying cockroaches and taking care to avoid Gordo, the poisonous cane toad who made his home under my bunk, I selected a pair of lightweight pants and a long-sleeve shirt from my backpack and shook them vigorously to make sure no scorpions were hiding in their folds. Thus assured that I would be the only occupant of my clothing, I dressed in the darkness. I slid my feet into a pair of strappy sandals, and my outfit was complete. Prepared for a night of surveying, I stepped outside, stood on the weatherworn wooden steps of the bunkhouse, and looked toward the center of our camp.

Five of my colleagues also were preparing for the night's events. Women and men from around the world, we'd gathered in this remote research base at the edge of the rainforest, drawn together by our interest in jaguars, sea turtles, and the complex interactions of these amazing tropical animals and their rainforest habitats.

Using rubber bands, we secured little pieces of red plastic to our headlamp lenses so that our bright white lights wouldn't disturb the sea turtles when they came ashore to dig their nests. Adult turtles return to shore only during nesting—a dangerous time when the marine-dwelling reptiles seek out the beaches where they hatched years earlier and venture onto land. Their sleek swimming bodies are not designed for quick movements on shore, and the female turtles are therefore exposed and vulnerable to terrestrial predators as they undertake the time-consuming task of selecting the perfect nesting site, digging into the sand with their flippers, and then depositing one hundred to two hundred soft, white, round eggs into the hole where they will incubate for approximately two months. Every step of this process is critical to the reproductive cycle of the green sea turtle. By shading our white headlamps with red plastic, our research group ensured that any turtles we encountered would have a dark nighttime environment in which to nest.

With my backpack stocked with water, snacks, and a notebook for recording data, I joined the other biologists in the center of the rustic camp. The darkened rainforest hummed softly beyond the split-rail fence. It was a beautiful, clear night, perfect for surveying. By the red glow of our headlamps, we quietly discussed the plan of action and plotted our survey transects.

Lined out in a row perpendicular to the shoreline, we would walk northward along the beach, carefully scanning for recent turtle tracks, freshly turned sand, activity in the water, or, most exciting of all, live turtles. If we encountered any of these, we would carefully record the location, time, size of the animal, number of eggs laid, and other important information. If we failed to spot any evidence of nesting, we would turn around at the end of the established transect and begin walking southward, back to the starting point, all the while searching for these important signs of activity. We would repeat these efforts until the rising sun tinted the salty horizon with a pink blush of sunrise.

With the survey details established, I pushed open the creaking, rickety gate and stepped beyond the safety of camp. My colleagues followed. In a silent line of bobbing lights, we made our way along the well-worn trail that led from our camp to the beach. As we passed under towering palm trees, our hands swatted at the moths that fluttered around our headlamps. Our eyes were trained on the ground to avoid surprising tarantulas or snakes, but we continually glanced up to scan the surrounding rainforest, hoping to catch a glimpse of the most elusive sight of all—one of the several majestic jaguars whose territories overlapped the survey area. Though subdued and dark in the deep hours of the night, the forest teemed with life.

After ten minutes of walking, we reached the end of the path, emerged from the rainforest, and beheld the moonlit Caribbean Sea. Small waves gently lapped at the sand; the inviting sounds of the surf beckoned our group onto the beach. Time to survey!

We lined up and turned our headlamps off; the light of the moon and stars shone brightly enough that everyone could see. If we encountered interesting turtle activity, we would turn on our red-tinted headlamps for closer examination. We began our slow march. I took the position closest to the sea for the first transect. The water lapped softly on my right-hand side, and I diligently scanned the waters and shore. I reminded myself to relax and breathe. The thought of encountering a turtle was so exciting that my shoulders kept tensing up automatically. The night was going to be long, and beach hiking was hard work. No need to expend extra energy on tense muscles.

My sandals flapped quietly on the wet sand as I kept pace with the line. Nobody spoke for many minutes. Who would be the first to break the silence? What would the lucky person excitedly announce to the group? Would the biologist from Australia stop abruptly as she pointed out the signs of a fresh nest? Would the researcher from Texas gasp as he spotted tracks in the sand? Would I, a Pacific Northwest native and recent graduate, be the lucky one to first lay eyes on a female green sea turtle splashing out of the surf?

Nothing. We reached the end of the first transect having encountered not a single thing. My colleagues and I switched positions as we turned around and prepared to walk back down the beach. I soon discovered that my original position had been both a blessing and a curse: I had enjoyed the benefit of the easy walking conditions at the beginning of my shift. Now relegated to the middle of the row, well past the firm sand bordering the lapping surf, I had to expend much more effort walking as I sank into the dry, deep sand.

For the rest of the night, I would walk the remaining transects in this soft sand, which would grow increasingly exhausting to navigate as the night wore on.

As the first hour came and went, my eager anticipation began to fade, and I no longer needed to relax my shoulders. The beauty of the Costa Rican nightscape faded into the background as we slogged through the endless sand. The details of the adjacent forest became a monotonous blur of foliage on the periphery of my vision. Failing to find any of the turtle signs I was so excited to see, I stared perfunctorily at the beach vanishing beneath my aching feet as though I were walking on a giant, sand-covered treadmill.

The group arrived back where we had started. Our second transect was complete, and it was time for a twenty-minute rest.

I lay back in the sand, resting my head on my backpack. My mind wandered, and I recalled that the survey crews of the last several nights had documented nothing during their five-hour shifts. I hoped that I wouldn't suffer the same fate. No turtles, no jaguars, and never-ending miles of nighttime beach hiking made for long, exhausting, boring work, especially after having slept for only a few hours beforehand. Speaking of which . . . closing my eyes felt so good . . . and the sounds of the wind and the sea were so soothing . . .

I grumbled as a colleague abruptly shook my shoulder. "Time to get going," he announced as I sat upright. For the third transect, I was stationed second closest to the edge of the rainforest. The soft sand slid in and out between my toes as my feet carried me northward.

It was shaping up to be a long night. The third transect appeared to be as much of a slog as the first two, until . . . "TURTLE! There! In the waves!" The identity of the lucky spotter went unnoticed as I eagerly turned in the direction of the pointing fingers. Even from many yards away, she was beautiful. The moonlight glinted off her plated back, sending sparkles into the inky water as the remnants of the sea ran down her carapace. Her long flippered feet left rake marks as she gripped the sand and hauled herself out of the buoying surf and onto the beach.

The green sea turtle paused at the edge of the water. Her large dark eyes appeared to scan the area, pausing only briefly on the group of stunned researchers standing several yards away.

"Keep your lights off!" came the whisper. "Remember, let her lay her eggs first. We'll measure her after she's done." Our top priority was to minimize stress to the turtle by allowing her space and safety to dig her nest and lay her eggs. Transfixed, we kept a respectful distance from the lumbering reptile. With my heart in my throat, I watched as the beautiful animal—with her expressive eyes, patterned flippers, and elegant shell—clawed her way up the beach toward the tree line. The turtle sounded distinctly bovine as she gasped and blew out her breath with the effort of her endeavor.

After making it three-quarters of the way up the beach, the turtle began to turn toward us. *What was she doing?* She continued to haul herself along in a semicircle, eventually facing back toward the sea. *Had she decided this wasn't the correct spot for her eggs to incubate and hatch? Did the presence of people frighten her? We were many yards away. Did something else scare her off?*

I scanned the tree line, searching eagerly for a hiding jaguar. The cats were known to crouch in the shoreline vegetation, lying in wait for an easy turtle meal. Did the turtle sense one of the large felines and decide it would be safer to nest on a different night?

I didn't see anything obvious that prevented the turtle from nesting, but she clearly did. Huffing and puffing, she pulled herself back toward the Caribbean.

"Lights on!" came a whisper. A red glow illuminated the beach as we jogged toward the turtle and surrounded her. She snorted and tried to heave herself toward the water. Intercepting her, the Australian biologist halted the reptile's progress by gently pushing her hands down on the smooth snout. I grabbed the measuring tape and laid it lengthwise along the carapace—nearly forty-six inches! Someone else recorded the measurement.

We examined the turtle to collect additional data. Although we had no way of accurately weighing her, we estimated she was approximately three hundred pounds, an average weight for a healthy adult female green sea turtle. Working quickly, we searched her flippers and body but found no identifying tags—no sign that this turtle had ever previously been encountered by biologists. The lead researcher pulled a small case out of his backpack that contained equipment to mark this new individual: a metal tag with a unique identifying number prestamped on its surface and a pair of clamps.

"Hold her steady! Grab her front flipper and hold it steady," he instructed me. I quickly complied. With the Australian biologist holding the turtle's head still, I had an easy time keeping the flipper in position. The lead researcher deftly inserted the metal tag into the clamps, wedged it over the edge of the turtle's foot, and clamped down sharply. The metal tag bent in on itself as it was secured into the flipper. I inwardly winced at the unsolicited body piercing, but the turtle didn't appear to mind—perhaps she was too distracted by the flurry of activity and bobbing red lights.

"Got it!" the researcher proclaimed, and he pulled the empty clamps off the flipper. "She's all recorded. Let her go now. We'll know if we come across her again, thanks to the tag."

Our mission completed, we backed away to allow the turtle space to recover and continue back to the sea. But just as everything seemed successfully wrapped up, a harsh, urgent whisper suddenly ripped through the darkness.

"LIGHTS OFF!" a voice hissed to my right. "Look, there! Be quiet and DON'T MOVE. Keep those lights off."

The urgency in the voice sent a pulse of fear through the group. We normally worked well as a team, so there was rarely a need for anyone to speak harshly or reprimand a group member. *What was going on?*

The biologist from Texas pointed. We looked northward and slightly off the coast, into the dark waters beyond. Were those . . . lights?

"POACHERS. Get in the bushes!"

As I scrambled toward the safety of the forest, my thoughts raced back to the turtle, abandoned on the beach. *Would she make it to the water? Would the poachers intercept her on her laborious journey back to the sea?*

We crouched in a row at the edge of the rainforest. Our own safety was potentially at stake. There was nothing we could do other than silently—and helplessly—watch the scene unfold.

A few hundred feet offshore, a small boat rocked in the waves. Bright white spotlights shone from its bow, piercing the darkness and sweeping up and down the beach. The poachers were looking for turtles on shore. If they spotted any, they would beach the boat and surround their victim. The turtle would be dispatched, its body harvested for goods that could be exchanged for much-needed money. The turtle parts would be sold on both the black market and in legal markets in Central America and beyond. The few regulations that existed to protect endangered species were hardly enforced amid the poverty, lack of government funding, and other social issues that plagued this part of the world.

In fact, during the long, bumpy bus trip from San José to Caño Blanco—the ride that had taken me from the Costa Rican capital to the outskirts of the rainforest and the beginning of my research adventure—a local man had boarded the rickety bus at a rural stop, hawking his wares.

"Huevos de tortuga! Huevos de tortuga!" he had shouted to the bus, simultaneously waving a damp plastic bag full of round white balls over his head. In the humidity, beads of moisture had flown off the bag and splattered on the grimy floorboards.

Turtle eggs. Sea turtle eggs, to be precise. Even though it was illegal to sell them, the bus riders had hardly looked up at the hawker's cries. To my dismay, I realized that the very species I had traveled so far to study and protect was being sold as a bus snack, ready to be washed down with a cold soda by thirsty travelers. They were so easy to buy—and for such a small price.

The poachers' boat bobbed in the waves as their bright lights continued to sweep over the beach. I knew this was where those sea turtle eggs had been harvested just a few weeks before. Poachers such as these dug up the nests of unsuspecting turtles and stole the eggs. The thieves carved the shells off the turtles' bodies and sold them as raw material for trinkets and jewelry.

My colleagues were well aware of this and probably knew more than me about the depressing situation faced by sea turtles and other endangered species. A somber sadness permeated the air as we watched, fearful of what the poachers might find. Confronting them was out of the question since guns and other weapons were tools of their trade. The group's safety was our top priority . . . even over that of the turtle that was still making her way to the water.

"Please . . . please make it back," I silently begged. I could barely make out the dark mass of the reptile lumbering toward the water's edge. Belatedly, I realized that the monotony and boredom we'd experienced while surveying was actually a saving grace in this situation—a lack of turtles to be surveyed and tagged meant a lack of turtles for the poachers to find and illegally harvest.

After what felt like an eternity, our lone female splashed through the surf, escaping the poachers' glaring lights. I breathed a sigh of relief as she slipped into the warm Caribbean waters and vanished into the night.

Having failed to illuminate any turtles on our stretch of beach, the poachers motored northward to seek their fortunes elsewhere, their lights shifting farther up the coastline. We began to relax, and the sound of blood pumping in my ears slowly subsided, giving way to the peaceful buzzing of the rainforest and the lapping of water on sand. When we finally spoke, our group made an executive decision to abandon the rest of the night's survey. Conducting another transect was too risky as we'd be going in the direction the poachers had headed. Plus, we had no way of knowing whether they would eventually double back toward us.

A subdued group of biologists headed back to camp. Some succumbed to their sleep deprivation and stared at their feet with glassy eyes as they trudged along the beach to the trailhead. Others were lost in thought, quietly processing the night's events. I fell into the latter group. At the trailhead, I paused, turning back to the sea before stepping onto the rainforest path.

Faint glimmers of color reminiscent of tropical fruits in their peachy hues were barely visible on the horizon. The colors faded upward, giving way to the black, star-strewn sky. Many hours remained before breakfast and the duties of the coming day, but the promise of sunlight made the night sky blush. In the distance, clouds rolled in over the sea; it would be another hot and humid day, filled with morning sunshine followed by the inevitable midafternoon rainstorm.

Today would also be filled with additional fieldwork. At 10 a.m., I was scheduled to survey for jaguars in the rainforest. Searching the forest floor for the tracks of this elegant predator was one of my favorite tasks. After that, I would have a few free hours to eat a simple lunch of beans and rice in the kitchen shack and then maybe take a book down to the beachside hammocks for some midday relaxation. I might lie in the noontime surf and soak my sunburned limbs, letting schools of small silver fish tickle my feet until the rain began to fall. I would get back to work later in the afternoon—data entry, perhaps? All of the researchers took turns with the never-ending task of digitally transcribing the group's observations and findings. I would confirm my assigned duties later, on the schedule posted in the kitchen.

First, though, came sleep. One last glance at the sea revealed nothing but the reflections of the stars.

But beneath those tiny points of light, a green sea turtle swam through the dark waters bearing the future of her species. Another night, another time, she would have another opportunity to give the next generation of turtles their chance at life. On that future night, maybe another group of biologists would shield their headlamps with red film, creating a gentle glow as they performed their work to understand and protect her species. Or perhaps glaring white spotlights would again sweep across the sand, as poachers sought out the turtle for the goods they could harvest from her body. Or maybe, just maybe, the turtle—guided by moonlight and some unknowable instinct—would find a dark, quiet beach free of artificial lights of any kind, on which she could nest and return to the sea undisturbed, as her ancestors had done for millennia.

Turning away from the stars twinkling over the Caribbean Sea and Central America's remote beaches, I headed back to my bunk.

Tyler Humphrey

Elizabeth Torrey lives and works in the beautiful shrubsteppe landscapes of the Pacific Northwest. She is a Certified Wildlife Biologist with a bachelor's degree in biology and a master's in wildlife science, and is completing a doctorate in leadership. Her experience as a professional biologist in both the public and private sectors has given her a well-rounded perspective on the interplay between human dynamics and wildlife conservation.

Elizabeth's love of all things wild has taken her across the globe in search of adventure. She has listened to mouse-tailed bats echolocating in the ruins of Luxor, observed torrent ducks diving into the white waters around Machu Picchu, protected her backpack from the marauding macaques of Angkor Wat, and hiked among Komodo dragons in their native habitat on the islands of Indonesia. When not working, studying, or traveling, Elizabeth can be found volunteering for her local animal rescue or riding her horse.

Tar Sands Tears

Laura Brodey

The water bodies I work around as a wetland field ecologist are mostly secluded areas with thriving wildlife. Elegant grebes bob along the water's surface, eagles peruse the landscape from lofty perches, and beavers slap the water with their tails. One early summer morning, I was headed to just such a wetland in northern Alberta, in an area farther north than I had ever been, with a member of our field team with whom I'd never worked. Our truck was loaded with an all-terrain vehicle (ATV) for each of us.

It was a gray morning. Mist hung low and color still seemed to be asleep. But in our warm truck, Derrek and I bantered and joked while taking in the landscape that moved past us. Looking ahead, I noticed an unexpected opening along one side of the forested road—a curious break in the trees lining the highway. As we approached, what caused this gap became clear. An expansive black tailings pond stretched to the horizon, bordered on its far end by a tall, belching smokestack. It was a dismal sight.

Northern Alberta is a land of contrasts. You can find yourself in the middle of remote, untouched wilderness where it is not uncommon to spot bears, wolves, and even cougars. I've come face-to-face with elusive lynx and regal great gray owls, along with other charismatic wildlife. But northern Alberta also contains areas of absolute devastation where pristine wilderness meets oil and gas development. Tar sands oil mining and its resultant lakes of black sludge are splattered across the region. These toxic tailings ponds are where the soupy by-products of oil extraction end up.

As I stared at the devastation, I was overcome by a crippling sadness and felt a sinking despair in the pit of my stomach. I tried to breathe through the unexpected emotion that assailed me as I gazed upon this travesty—a sight far more shocking when seen in person than in photos or videos. I felt an overwhelming sorrow for all the animals and trees that had been wiped out and replaced by this filthy, toxic brew that would afflict the land and water for decades (if not centuries) to come. I recalled that I had recently met a fellow whose job it was to prevent the ducks that landed in Alberta's tailings ponds from drowning in the sludgy abyss. He had to grab the ducks, twist their necks, and dispose of them. I later learned that more than sixteen hundred ducks died in just one company's Albertan waste ponds in one year. (As many as a million birds perish each year in similar tailings ponds in the US.)

Derrek seemed to share my disgust, expounding animatedly about the impossibility of restoring such polluted and degraded landscapes to anything close to natural ecosystems, despite the claims by the oil and gas industry that it could do so. As I listened to him, I fought back the tears that suddenly welled up in my eyes. In my mind, all I could see were faceless businessmen sitting around a boardroom table making cavalier decisions that would fill their wallets but irrevocably change Canada's natural landscapes.

As Derrek elaborated on the futility of tar sands restoration projects, it became more and more difficult for me to escape the vivid images that flooded my mind. By the time he looked over at me, tears were pouring down my cheeks. Belatedly realizing the effect that his words were having on me, he apologized profusely as we drove on.

The rest of the drive to our field site was tensely quiet as I tried to regain my composure, along with a measure of hopefulness. I paid scant attention to my surroundings as we continued to our destination. But I couldn't help noticing roadside displays of massive oil extraction machinery that were unabashedly exhibited as if they were a point of pride. Rather than being impressed, all I felt was a profound disappointment in humanity and its self-serving plunder of the natural world. At the same time, the irony didn't escape me that our truck and ATVs ran on gasoline and that we mainly used oil-and-gas exploratory cutlines to travel through the forests to our remote field sites.

When we finally reached our destination and Derrek parked our truck in a grassy patch, I quickly made an excuse to be by myself for a moment. Once I was alone, my tears flowed even more freely than they had earlier. I crouched on the ground, sobbing, then paced back and forth, wiping away still more tears. Eventually, though, the natural surroundings soothed me, and taking deep breaths, I pulled myself together and found the strength to resume my work.

Arriving back at the truck, I discovered that Derrek had offloaded the ATVs all by himself. While not the safest course of action, his small act of kindness reminded me of humanity's dual nature—its selfishness and its selflessness. And it made me think that maybe, just maybe, there are enough caring and giving people on our planet to tip the scales and allow us one day to find a better balance that is less destructive to the natural world.

As we drove the ATVs to our wetland field site, we got a little bit lost in thickets of vegetation and spent some time stuck in mud, but paddling around all day in a pristine pond surrounded by lush aromatic plants and cheerful birdsong was thoroughly restorative.

Since my job is fairly unconventional, I often find myself trying to explain my field biology work to perplexed individuals outside of my profession. After all, my chosen field is rarely lucrative and doesn't always provide stable employment. The most frequent response to my job description is a blank stare. My occupation is so far removed from most North Americans' conception of work that many people struggle to visualize what being a field biologist really means. More often than not, I have to wake up before sunrise, dress for weather extremes, drive or hike to the middle of nowhere, carry heavy gear, find discreet ways of going to the bathroom outdoors, battle biting insects, avoid poisonous plants, identify an endless array of creatures and vegetation, race against a setting sun to take repetitive measurements, and record everything on paper that I pray won't get so wet that my data become illegible. Needless to say, I am not surprised when my audience looks at me with a mixture of bafflement and pity as I tell them what I do.

But while we field biologists are arguably a bit off our rockers, what is most important to know about us is that we are here, working tirelessly to document, catalog, and understand rare and beautiful ecosystems in an effort to give them every chance of surviving humanity's destructive ways so that they may flourish into the future. We are here, caring about all living things—from lichens to loons, from otters to old-growth—and illuminating the delicate interconnections among them. The data we gather are instrumental to conservation efforts. And through our work, we give ourselves hope for tomorrow—and perhaps provide a little hope for the rest of humanity, too.

Alexandra Janes

Laura Brodey has followed a lifelong calling to connect with—and protect—the natural world. As a child, she could often be found high in the branches of a tree or feeding insects, lost in quiet observation of the life around her. That early wonder evolved into a mission: to blend her deep sense of stewardship for the planet with hands-on environmental work. She holds a bachelor's degree in environmental studies, certificates in ecosystem restoration and digital mapping, and a master's in forest conservation.

Laura's boots-on-the-ground career has spanned stream restoration, bird banding, wetland and plant surveys, urban forestry, soil science, and more, and it has carried her across Canada, to the jungles of Costa Rica, and to the heart of Southeast Asia. Whether she's documenting insect behavior or befriending a majestic deer, she finds magic in every wildlife encounter. She believes the future depends on rekindling our reverence for the wild—and she's doing her part to make that happen.

The Little Red Bat

Lindsey Zarecky

It's a hot, humid summer night in North Carolina—perfect for fieldwork. Four of us—once just colleagues but now friends—have spent the past few hours preparing for a long night of catching bats. For us, summer means surveying season, when we capture bats to investigate which species live in our area and assess the bats' health.

Laughing, cursing, and genuinely having a good time, we have strung four of our fine-mesh mist nets over waterways and across walking trails in the woods, hoping for a good night of bat catching. With any luck we will capture at least half a dozen bats. We hope to get multiple species, males and females, adults and juveniles, because tonight is no ordinary night. Tonight's work is special; unlike our usual surveys, this one has an audience. Twenty rambunctious, giggling, wide-eyed summer campers swarm around our field site, eagerly awaiting close-up views of bats. So tonight, we are not just bat biologists but also educators.

The sun falls below the horizon. It's go time! We put on scrubs, waders, boots, leather gloves, latex gloves, hats, facemasks, and headlamps. Dressed in our layers of personal protective equipment, we arrange our data collection station and set our timers. Every eight minutes, one of us will walk to each of the nets and shine our light over it, searching for any bats that are caught in the mesh. As alarms go off, we take turns checking the nets. Campers fidget around our little makeshift data station, hoping someone will return with a bat.

"I got one," Olivia calls out. The campers' eyes widen and they squirm even more as their excitement mounts. Their first real live bat! We watch the light from Olivia's headlamp move closer as she approaches our eagerly awaiting group, carefully carrying a soft mesh bag containing the bat she extricated from the net. The campers jostle around the data station, positioning themselves for the best view of the elusive creature.

Reaching into the bag, Olivia pulls out a small bat. The kids eagerly lean forward, squealing with excitement as she holds up the bat for them to see. Olivia then sits down, bat secured in her gloved hand, and we begin our routine. While Rada prepares to hand Olivia the measuring equipment, I grab the data sheet and pencil to record data.

Gently but skillfully, Olivia examines the bat and calls out information for me to record. "It's an eastern red bat," Olivia notes first, identifying the species based on its orangish coat. "It's a female, but she's not lactating so isn't caring for young at the moment." Using a ruler, Olivia measures the bat's forearm length. Next, she extends the bat's wing while Melissa shines her light on it. We can't see through the finger joints, so we know that her bones have fully fused, indicating that she's an adult. As we work, the campers eye the little animal, exclaiming:

"Oh my goodness, it's so cute!"

"It doesn't look scary at all."

"Aww, she's adorable."

"It's so tiny!"

After we've taken our measurements, we gently place the bat back into the bag and use a handheld scale to weigh bat and bag together.

Rada explains to the campers that bats in North Carolina can acquire white-nose syndrome, an often-deadly introduced fungus that appears on areas of bare skin, like a bat's snout and wings. The irritating white fungus, which has killed millions of North America's bats since 2006, wakes hibernating bats in the winter. Because the bats' metabolism increases when they wake up, they can die from starvation since there are no insects to eat. Carefully removing the bat from the mesh bag once again, Olivia holds it in her hand for another brief moment, giving the campers a last view of the tiny creature while also looking for any scars left by the fungus. Fortunately, this bat appears healthy.

Then comes the big moment. Olivia leads the way into a clearing with the bat in her hand. As the campers follow behind her, she explains what is going to happen. Pausing, she raises her arm and we hold our collective breath, waiting for this beautiful animal to take flight. The little bat takes off from her hand . . . and immediately falls to the ground. Olivia gently retrieves it, then raises her arm again, giving the bat another opportunity to fly. But again the bat flutters weakly and falls to the ground.

Nervously, Olivia turns to me and whispers, "She won't fly and I'm not sure why. Can you take her?" Concerned, the kids ask what's happening. After calmly explaining that the bat may be stressed and needs to be taken into the woods to rest, Olivia redirects the kids' attention back to the data station to study another bat, while I walk away with the stricken red bat in my gloved hand.

The bat is breathing heavily but shows no signs of any injury from having been tangled in the mist net. Nevertheless, her rapid breathing and unwillingness to fly suggest that she is severely stressed. Rada joins me and together we offer the bat some water. I hold her gently in my hand for a moment, hoping she will recover and return to the sky. But she remains motionless. Anxious to ease her stress, we take her deeper into the woods, farther away from the kids, and place her on a tree branch. A few moments later, though, she is on the ground again. The bat's prospects look dire, and the longer we are in the woods, the more the kids' anxiety grows.

After a quick discussion, Rada and I decide that a more serious intervention is required to revive the little bat. Based on her experience with rodents, Rada suggests we try sugar water, if we have any. When small animals with high metabolisms are stressed or their energy reserves are low, sugar water is just the thing to give them a boost. With the little red bat safely tucked into the mesh bag, we return to the truck and dig into our field bags but find no sugar and no syringes. It's time to get creative.

Looking at her midnight snack of freshly diced cantaloupe, Rada says, "I have an idea—a crazy idea, but an idea." She opens a water bottle and begins to squeeze sweet, sugary cantaloupe juice into it. We tip the bottle to the little bat's mouth and sloppily feed her the sugary water.

Over the next ten minutes the bat visibly begins to regain energy, moving her wings and tilting her head toward the bottle. We return to the data station with the bat and juice bottle and continue to offer her juice for the next half hour. With each sip of juice, the bat gets stronger. As her feistiness returns, she attempts to bite my fingertips and squirm out of my hand. We place the now-sticky bat upright on a nearby tree branch. Ten minutes later, she turns herself upside down and hangs by her feet in a comfortable bat-roosting position. Another ten minutes go by, and finally she takes flight.

In that moment, the four biologists' excitement rivals that of the campers. We giggle alongside the kids and continue talking with them about bats and protecting wildlife. Once the night's netting session has come to an end and the campers have nestled into their tents, we can hear them chattering about their awesome experience and exclaiming about how amazing bats are.

Of all the nights we spent mist netting bats, it was *that* night—with an audience of impressionable youngsters—when we were reminded how vulnerable small mammals can be. But the little red bat showed us—and the kids—that we can help animals that are in trouble. Logic, quick thinking, resourcefulness, and our innate desire to help creatures in need, particularly when our work has inadvertently caused them distress, are some of the best tools we biologists have. But all of our field vehicles are now outfitted with sugar water and syringes—just in case!

Hannah Lombardi

Lindsey Zarecky grew up in a small town in southwestern Pennsylvania, where her family fostered her love of the outdoors. Camping, hiking, swimming, and playing outside were favorite pastimes starting in childhood and continuing into adulthood. She moved to North Carolina to get her master's degree at University of North Carolina Greensboro, where she studied bats. Upon graduating and joining the workforce, she saw firsthand how challenging and isolating the field of ecology can be for women. But with a knack for teaching, a love of science, and a longing to contribute to conservation, she didn't let any obstacles hold her back from her goals.

Lindsey is the vice president of conservation and research at the Greensboro Science Center, where she helps protect many species native to the US and the Caribbean.

True North

Kira A. Cassidy

Braced against a biting wind rushing in from the Arctic Ocean, I stood on a hill overlooking miles of tundra. I could see three herds of muskoxen in the distance and a family of arctic wolves just close enough for me to faintly hear their happy yips and whines. Behind me were several piles of gear—tents, all-terrain vehicles, duffel bags—packed up and waiting for the helicopter.

I hadn't showered in nearly seven weeks; my clothes were covered with dirt and now fit me loosely. I even had new freckles across my nose from spending so long in the twenty-four-hour polar sunlight. The journey home would begin in less than twelve hours, so I soaked in those last few moments on the tundra until it felt like I'd made an exchange with Ellesmere Island: here's part of my soul for part of yours. Even if I never saw this place again, it would always be part of me.

I had traveled to Ellesmere, a large island in the High Arctic only five hundred miles from the North Pole, because of a wildlife documentary. The production company wanted a biologist to accompany the filmmakers, and my experience in Yellowstone National Park studying wolves earned me the job—interpreting wolf behavior and offering my professional insight into wolf pack life. I was nervous those first days. *Was I the right person for this job, which wasn't clearly defined, in a place far from anywhere I'd ever been before?*

I relaxed a bit after our helicopter scouting mission to find wolves and a den. After several hours flying over the tundra, I noticed some green grass, then one wolf, and then a second one that stood up before slipping underground. A den! Within a day we had moved in next door, setting up our camp about half a mile from our new tundra-mates, the Mount Lockwood wolf pack. I was relieved to be in the field, and I was able to relax even more as my responsibilities became clearer.

Our four-person camp—producer, two cameramen, and me—was a happy place. Life revolved around the wolves, but we also cooked meals and played board games or told stories. It was close to the summer solstice when we arrived, so the sun made small circles high in the sky. It only started dropping closer to the horizon, swinging 360 degrees around camp, in the last week or two of filming. Because the sun never set, we took turns watching the wolves and followed along whenever they went hunting, no matter what time our watches said.

There were a lot of chores to do each day, and I found a rhythm monitoring the wolf den, recording behavior, joining the crew while filming, and maintaining camp. It could be tough work, and sometimes it seemed like most of my job consisted of hauling heavy items (water jugs, gas cans, tripods, cameras) from one place to the next. Everything was coated in fine dust, dirt, or thick, heavy mud. But the sun shone and the air smelled like salt water and icebergs. Muskoxen fur, shed in long strands, caught on tiny tundra plants. Lemon-yellow arctic poppies reached a few inches into the sky, and purple saxifrage blanketed entire hummocky hillsides.

As for the wolves, they mostly slept and hunted and played with their three pups. The adult wolves occasionally inspected our camp, and the beta female (second in the dominance hierarchy) even stole a sock I had left drying on my tent. I retrieved it quickly as she walked away, but not before she had ripped a few holes in the heel.

The wolves' family dramas played out right in front of me. I heard the whines of hungry pups and watched their mother spot a tiny speck of white two miles away, then get up, chase down a young arctic hare, and immediately bring the food back to them. I watched the oldest wolf dote on the pups—cleaning them and gently playing with them—but also saw her submissively tucking her tail and licking the faces of the other adults. The yearling male often left for days at a time, and his return sent the whole pack into a wagging, licking mass of happy bodies. It was truly a family of individuals with unique personalities and skills.

Details of the wolves and the bird calls and the smells and tastes of the Arctic etched themselves into my mind. And even though my muscles ached and I slipped exhausted into my sleeping bag each "night," I felt strong and centered and more aligned with my innermost self than I'd felt in my entire life. I'd found something unnamable in that wild place among the wild creatures. Which is why, just before leaving, I stood on that hilltop, feeling my heart cradled by my surroundings. The ground supported me from below, the sky gently protected my head, and the hills embraced me on all sides. And so I made that pact with the island: part of me will stay here forever.

Returning home to Montana was easy in some ways. I was ready to take regular showers and eat salads and avocados and wear only one layer of clothing for a while. Mostly I think it was easy because part of me was still up in the Arctic. Not in a way that felt like I was split or somehow divided, but more like my spirit had grown and expanded. Ellesmere had become a part of me, and the growth I had experienced there gave me strength and clarity, which helped me feel reinvigorated in my job and in my healthy friendships. Over the next year, I also met a man named Conall, fell deeply in love, and slowly started opening up to him.

Two years went by and the subject of Ellesmere sometimes came up when I shared nature stories or talked about the trials of expedition camps. That place was still living in my core. It wasn't easy to explain how I felt like I could turn my thoughts inward and reach with my mind and heart to that piece of me living among different mountains. So I didn't really try to explain it to anyone; I cherished that feeling of deep connection to Ellesmere and also relished my Montana life. I occasionally thought about what it might be like to go back to the Arctic, but I was also content that I'd experienced it at all.

Then, one day while I was starting to plan a road trip with Conall, I received an email from a different film production company: they were going to Ellesmere and hoping to film arctic wolves, their team's consulting biologist had canceled, and they wanted to know if there was any way I could join the expedition. In *four* weeks. I was thrilled, but I'd scheduled time off from work for the road trip, and canceling with my partner was disappointing. He traveled for his work, so finding blocks of time to do things together was rare. I took a risk and accepted the job on Ellesmere on the condition that they add Conall to the team. The production company agreed, and I switched from planning a road trip north to planning an expedition north of north.

Doubts crept into my mind: *What if I went back and the place or the experience was totally different? Would we be able to find any wolves at all? Would the place still feel like home?* I also felt vulnerable and exposed. Ellesmere had become such a core part of my identity that sharing it with Conall felt scary. But this was a person I adored, a person I'd shared my mind and heart and body with for two years.

He was also the one person who seemed to understand my connection to Ellesmere, who truly saw the delight in my eyes and heard the passion in my voice when I spoke about it. So the idea of sharing my innermost connection to my tundra home with him also felt brave. Conall and I were so alike in many ways that I decided to lean in and open up even more. I just knew that he, too, would fall in love with that wild, frozen world.

That second trip to Ellesmere was beyond any of my dreams. Since I was the only person on the crew who had been there before, I took on much of the planning and logistics. Where my first trip had started with some self-doubt about whether I could be a valuable team member, this time I knew that I was. We set up camp at a different pack's den site. The Eureka pack had six adults and three pups. The pups were about five weeks old—two brothers (one light tan and one dark brown) and one little sister. Some of the adults were curious about us, and after initial, tentative sniffs of our gear, they accepted our presence as observers in their backyard.

Being back in the land where I had grown and learned so much about myself was invigorating. I collected scientific data on wolf prey, den use, pup growth, travel patterns, and individual personalities. The two filmmakers were often busy with their equipment. They used drones and small cameras designed to look like fake rocks that could be driven slowly over the tundra by remote control, allowing close-up glimpses of the wolf family.

Conall and I helped the crew, and we also used all-terrain vehicles to explore and find routes to follow the wolves as they traveled. We investigated locations downloaded from radio collars that a few of the wolves had worn for the previous year. Usually these locations were small hills affording sweeping views of the wolves' territory and great places for a wolf (or human!) to nap, or low, wet ravines with muskoxen skeletons that had settled into thick mud. Often the carcass had been picked clean months earlier, allowing us to imagine what life was like for the wolves during the four solid months of winter darkness.

We got to know the place intimately and gave names to the creeks and valleys and mountains. We got wet and muddy and covered in mosquito bites. We bathed using water from tundra potholes with a permafrost base. We howled with a wolf pup when it got lost. We sat and watched quietly, wrapped around each other, as the whole pack played in amber light. We laughed and talked and slept under the circular motion of a late-summer polar sun.

If my first trip to Ellesmere made me feel like I had found my purpose and forged a special connection to a place, my second trip felt like I was bringing my loved one into my richly woven bond to Ellesmere, and in doing so, inviting him to get to know a part of me I'd only ever kept for myself. Sharing so much wasn't always easy. I felt emotionally bare, and my openness felt full of risk. Sometimes I wondered: *Does he realize how special this place is? And what it means that I'm sharing it with him? Does he know I risked my chance to return so he could come too?* Other times I could tell by the way he looked at me or held me that he was in awe of the place and intensely grateful to me for crafting this shared adventure.

At times it felt like we had traveled together to another planet, but one that felt like home. I couldn't dream up a more romantic, soul-testing, and energizing fusion of our partnership. We had traveled to a corner of the world few even knew existed and lived in a remote area seen by only a handful of other humans. Our neighbors were some of the most fascinating wild creatures on the planet and were completely unafraid, even inquisitive, around us. I collected data and interpreted wolf behavior for the film crew, and both the documentary and scientific endeavors were successful. As was the profound sharing of my most special place with the person I loved.

Leaving Ellesmere that second time was very different from the first. I still stood alone on a hilltop to say my goodbyes, but this time I projected out my thanks to Ellesmere and also made a promise to come back. Conall and I had already started discussing plans to return for our own larger project, one that spanned nearly the entire six-month-long day—sunrise in late February to twilight in September. With every planning session I could feel my heart rejoice. When we returned home, I started to coordinate with several other scientists to initiate a large research project, and Conall began to organize his media contacts to document the project and the lives of arctic wolves in a magazine article and film. For nearly a year, we worked on plans to return to what had become *our* place.

It's massively confusing when a person you love pulls away. At first I thought stress was causing Conall's withdrawal, but eventually, with a mixture of sadness and panic in his eyes, he told me he wanted to end our relationship. I tried to understand and patiently listen, hoping these were just fleeting thoughts, but over time, it became clear that our relationship was irreparably damaged. Confused and hurt, we stopped communicating, and my heart shattered. Then, when Conall continued planning the big project and returned to Ellesmere without me, I felt like my soul had been violently stolen, abused, and discarded.

Since then I have struggled to understand what Ellesmere means to me now. Part of me still lives there, and although that piece of me is aching and bruised, it will forever reside in that wild place. Some days the pain and betrayal completely cloud over my good memories of Ellesmere. But I also know that in sharing something so dear, I took a risk and made a brave choice. I chose to love and live openly and to wholeheartedly share the most precious parts of myself. My decision left me vulnerable, as brave choices often do. They leave an opening for pain or manipulation or carelessness by others. But such choices can also foster life-changing connections with people and places, connections that have profound depth and meaning.

Maybe I'll return to Ellesmere someday. Or maybe all the memories I made during my adventures there will last me a lifetime. I only hope that on the hard days, the memory of a midnight sun finds a few cracks in my grief. I hope it shines through to light up the images in my mind of fluffy white cottongrass flowers, a sawtooth row of glacier-bound mountains, and a whole family of arctic wolves, smiling and wagging and playing in golden light. And

I hope that no matter where I am on Earth, my soul points true north, toward the piece of me that still lives amid the ice and the howls.

Ronan Donovan

Kira Cassidy grew up in Illinois watching white-tailed deer and wood ducks. After finishing a degree in natural resources, she moved west to Yellowstone National Park. For nearly two decades, she has worked for the Yellowstone Wolf, Cougar, and Elk Project, rising from volunteer to field technician to research associate. Her master's research at the University of Minnesota focused on wolf territorial behavior in Yellowstone, and her current research projects focus on wolf-pack social behavior and how pack dynamics compare to behaviors seen in other social mammals, including humans.

Kira has also worked and consulted on the ecology and behavior of other wildlife, including arctic wolves on Ellesmere Island and desert bighorn sheep in Utah's canyonlands. In addition to conducting and publishing research, she communicates science to lay audiences through writing, public speaking, and watercolor and digital art. Kira lives along the Yellowstone River in Gardiner, Montana, with her hound dog, Toby, and enjoys hiking, painting, and reading. Her favorite days always involve a glimpse of a wild animal going about its daily life.

Fear on the Mountain

Sophie A. H. Osborn

The low drone of an approaching plane disrupted a tenuous morning silence that hovered over the arid landscape like a fragile veil. Seconds later, my two-way radio crackled.

"Osborn, this is Parish, do you copy?" the disembodied voice of our project director asked.

"10-4, I copy," I responded, using the universal acknowledgment of a radio transmission.

"Okay," Chris continued. "Our plane is almost above you now. We can see your truck."

"I'm glad you were able to get gate keys from the landowner so I could access this ranch," I said.

"Me, too," Chris agreed. "This place is remote! Why don't you give the road that you're on a try. It's heading in the right direction."

"10-4," I replied, depressing my truck's accelerator. The vehicle rolled forward and quickly gathered speed. Static from my radio-tracking receiver, which was tuned to the transmitter frequency of the young California condor we were looking for, competed with the buzz of the plane overhead.

Hatched in captivity in April 2001, Condor 240 had been released to the wild at northern Arizona's Vermilion Cliffs the following February as part of an effort to recover this critically endangered species. In the months following his release, 240 had shown us that even for a bird that is master of the skies, with a wingspan of more than nine feet, he was particularly driven to fly. Long after the sun had set each evening and the other condors had settled down to sleep, 240 would remain aloft, gliding back and forth over the release-area cliffs in the waning light. And rather than making short preliminary forays away from the release site like his fellow juveniles did, 240 had accompanied our most adventurous condor on a seventy-mile flight northwest to Zion National Park in southern Utah on one of his first trips post-release.

Impressed with his long flight and adventurous spirit, we were hardly surprised when he disappeared on his third visit to the Zion area in mid-August. Two years earlier, his older sister had gone missing in the same area, precipitating endless searches until we finally located the thriving youngster months later in a remote corner of Grand Canyon National Park.

Every captive-reared condor that was released to the wild wore an identifying number tag and a radio transmitter on each of its wings so biologists could track the bird's movements and be on hand to foster appropriate behavior when the bird first encountered people and developed areas. In addition, we continuously monitored all the reintroduced condors to determine what they were eating.

If we thought a condor had fed on an animal that had been shot, we captured the bird and tested its blood to determine if it had ingested lead. When a lead bullet is fired into an animal, it fragments, leaving hundreds of tiny metallic particles in the carcass that scavengers like condors can consume inadvertently. Whether condors survived after feeding on such animals depended on the amount of lead the birds had ingested and how quickly we could begin treating them. If we captured lead-poisoned condors in time, we could usually save them. But too often, the birds died. Historically, condors had a lifespan of sixty to seventy years. Today's condors were lucky to survive into their teens.

Worried that Condor 240 might have encountered a lead-filled carcass, members of our field crew had spent countless days driving alone all over southern Utah and northern Arizona searching for him. With a radio-tracking receiver by our side and a tall antenna on the roof of our truck, we listened to the static emanating from the receiver, straining to hear the faint *blip blip* that would indicate that 240 was in range.

After failing to find 240 during our ground searches, we had tried to track him from a plane. During a second flight, Chris had finally heard a faint transmitter signal coming from a sprawling ranch northwest of Zion National Park. Now, he had relocated the signal from the air and was directing my movements along rugged ranch roads so I could find 240. After passing through a locked gate, I sped over a rough road leading to a distant water tank. Racing along below the plane that shadowed me from above, I felt for a brief, surreal moment like I was acting out a part in a James Bond movie.

"Are you getting his signal?" Chris asked as I reached the end of the track.

"Negative," I responded.

"Okay, we'll try another road. We have another option." As the plane circled overhead, I backtracked, then took a different road heading west. Suddenly, I heard the almost imperceptible *blip blip* of one of 240's transmitters.

"I've got it!" I called up to the plane before jumping out of my truck to use a Yagi antenna, which allowed me to determine the signal's direction.

"10-4," Chris responded. "That's great. Good luck finding him! We'll head back to the airport now and will fly back over this afternoon if we don't hear from you."

Moments later, the plane's thrumming motor faded into the distance and I was enveloped in silence. I drove on a bit farther, parked, then shouldered my pack, readying myself for a prolonged hike into unknown territory. As I stepped away from my truck, I took a bearing with my compass (GPS units not yet being part of our field equipment), noting the direction in which I was headed in case I needed help finding my vehicle on my return journey.

Climbing steadily, I hiked up a steep, timbered ridge, following the direction of 240's faint and sporadic transmitter signal. After nearly an hour of winding my way through bushes and trees, I reached a heavily grazed meadow. A golden eagle soared over me, its head swiveling to pierce me with its fierce gaze. 240's signal was booming now, and as I approached a stand of dying aspens, I scanned the ghostly branches, hoping I would see his bright eyes peering down at me. But I searched the treetops in vain and soon resigned myself to scanning the ground. It wasn't long before I found him.

Condor 240's desiccated body lay beside a fallen log, amid scattered deadfall. His tail was pitched upward and his wings were spread out, as though he'd dropped out of the sky midflight. I stared intently at those great wings.

A shiver ran up my spine. Not a feather was out of place. But 240 no longer wore his identifying tags and transmitters. Somebody had removed them.

Eyes wide, I looked around, searching for 240's transmitters, searching for someone who might be watching me. Moving my Yagi antenna in a slow circle, I fiddled with the controls on my receiver, trying in vain to narrow down the location of 240's still-booming radio signal. There was no discernible difference in the signal's volume in any direction. I must have been within feet of the transmitter but saw no sign of it. Removing the antenna from the coaxial cable that attached it to my receiver, I held out the end of the cable and moved in another slow circle. This time, the signal was slightly stronger on my right side. I stared at the ground. Small branches littered the dirt, and as I looked at them, my mind racing, it occurred to me that their placement seemed unnatural, as though they had been laid down intentionally rather than falling haphazardly.

Pulling my camera out of my pack, I took several photos of the sticks and the bare ground under them. Then I moved the woody debris aside and dug into the dirt with my fingers. Within seconds, I felt 240's vinyl identification tags and their attached transmitters. Drawn unwittingly into a situation that suddenly felt intensely threatening, I again glanced around uneasily, searching for a hidden figure amid the scattered trees.

After reassuring myself that I was alone, I left the buried transmitters as I had found them, methodically photographed the crime scene, and recorded everything I had seen and done in my notes. I then hiked fifteen minutes or so to the top of a nearby ridge, hoping to get cell phone service. In luck, I called Chris so he could alert the authorities. Sitting on the ridgetop, arms wrapped around my knees, I gazed out at distant gray cliffs and a sparkling reservoir and settled down for the hours-long wait for Chris and the US Fish and Wildlife Service law enforcement agent.

As images of 240 filled my thoughts, a movement below the ridgeline caught my eye. A rider was approaching on horseback. A rider who probably didn't know that I had permission to be where I was. A rider who might have shot a critically endangered California condor. And buried the evidence. Adrenaline surging, I scrambled behind some bushes and dropped to the ground, heart in my throat, as the rider came fully into view, a hat pulled low over his eyes, a shotgun lying across his lap.

Heart racing, I pressed myself into the dry dirt, trying to make myself invisible as my mind flashed back to the only time I had ever felt such intense fear before. I'd been a teenage girl out for a morning run when two beer-drinking men had passed me in their low-slung sedan. Braking, they'd turned around and then passed me again, leering. Once again they'd turned, then pulled over, waiting for me to approach. I'd ducked into the yard of the only house for miles around and cowered behind garbage cans as the men drove into the yard, got out of the car, and searched for me before deciding I had gone into the house. Then, as now, I'd felt acutely vulnerable and had a surreal sense that I was outside my body, watching from afar the events that were unfolding around me. Events over which I had no control.

The rider neared. His horse's hooves kicked up small circles of dust. And then they moved past me. Fervently, I hoped the man wouldn't see me, hoped he didn't have a wide-ranging dog that would discover my prostrate form. After an eternity, the horse's plodding hoofbeats faded, and I began to breathe again. Later that afternoon, I hid once more as a wave of grazing sheep approached and then surrounded me, spooking and jumping away as they caught sight of me. But the flock and its unseen minder eventually moved on, and my pulse again returned to something approximating normal.

Hours later, darkness dropped silently around us as we somberly loaded 240's body into Chris's pack. Thanking me for my careful treatment of the crime scene, Special Agent Bell repacked her equipment.

"So, time to find our way back to our vehicles in the pitch dark," Chris laughed, injecting a much-needed moment of levity.

"I took a compass bearing," I said, spinning the dial 180 degrees to navigate back to my truck.

"Great. Lead us out of here, then."

As we trudged down the mountain, fighting our way through sudden tangles of brush in a moonless darkness that hid any possible landmarks, I fervently hoped that the compass skills I had learned when searching for goshawk nests in Idaho's forests years before would serve me well now. An hour later, we stumbled onto the road twenty feet from my truck, and I smiled for the first time that day.

In the months that followed, as we awaited the results of 240's necropsy and the conclusion of the federal investigation into his suspicious death, one of our four-year-old condors was felled by a hunter's arrow and left to bleed to death on the forest floor. Two months later, another young condor was shot by a hunter.

But when the US Fish and Wildlife Service finally relayed its findings to us regarding its investigation into Condor 240's death, we learned to our profound astonishment that he had *not* been shot, as we had assumed. Instead, 240 had been found dead by a ranch hand who, aware that condors were federally protected, panicked, convinced that he would be blamed and suffer severe consequences—perhaps even deportation—if anyone learned of the bird's death. Carefully removing the transmitters, he had attempted to bury the evidence. But unbeknownst to him, that evidence had remained trackable, and its burial had made him appear guilty. Nevertheless, the necropsy of 240's remains conducted by the forensics team revealed unequivocally that our young adventurer—like so many condors past and present—had died not by being shot with lead but by ingesting it.

Illegal shooting of condors was inarguably an ongoing problem. But as the coming years were to prove, it was lead poisoning that threatened the condor's continued existence. Lead poisoning that had caused the species' near extinction. Lead poisoning that undermined the bird's future recovery. In the course of reintroductions to save a species that had once numbered only twenty-two remaining birds, approximately two-thirds of adult and a quarter of juvenile condor deaths were attributable to the ingestion of lead.

As far as recovering endangered species goes, the fix is a relatively simple one. If hunters used widely available nonlead ammunition, condors could benefit from feeding on the animal remains but would not ingest lead. Countless golden eagles, bald eagles, ravens, and other scavengers would also avoid dying after ingesting lead fragments in carcasses—and people, who can inadvertently ingest lead particles when eating deer and elk shot with lead bullets, would similarly avoid the health hazards of consuming the dangerous metal. But while some hunters now use nonlead ammunition for hunting, others decry any mention of the perils of lead ammunition as a hoax or an infringement on their Second Amendment rights. And lead-filled carcasses continue to litter our landscapes.

Despite clambering around on perilous cliff ledges, hiking alone for hours at night in the Grand Canyon, and carrying calf carcasses (condor food) on my back after dark in areas with mountain lions, I never again felt acute fear while working with condors. But I often feared that my condor charges wouldn't make it. And I suffered from a low-level eco-anxiety—that latent, gnawing fear that we are losing all that makes the world around us so beautiful, inspiring, and uplifting. In the Anthropocene—the current geologic age when humanity is the dominant force impacting our environment and climate—such anxiety is prevalent among biologists.

Many women biologists face fears when working in the field: fears of being accosted by strange men; of falling off cliffs or down talus slopes; of being swept down raging streams, attacked by predators, or bitten by snakes; of succumbing to heatstroke, being struck by lightning, or getting lost in remote country. These fears can be acute—but they are temporary, and we overcome them. We must so that we can do our work. The deeper fears—of losing creatures we've worked to study and protect, of seeing habitats destroyed and populations plummet, of losing biodiversity and witnessing extinction—are more pervasive, more enduring, but we persevere, taking heart from wild places and inspiration from the wildlife we love.

Whenever I watched condors drift, with effortless grace, over vast landscapes that had seen their unmistakable shadows for tens of thousands of years, my fears subsided. Watching these great-winged messengers of hope, I found myself, instead, celebrating all the good that humanity can achieve by harnessing its collective might to protect and conserve rather than to exploit and destroy.

Author's note: This story is adapted from a passage in *Feather Trails: A Journey of Discovery Among Endangered Birds*.

Wild Friendships

Jungle Shoes

Holly Bradley, Masha Edmondson, Bruna Campos, Taylor McDowell, and Kate Derrick

"Give a girl the right shoes and she can conquer the world."
—source uncertain; misattributed to Marilyn Monroe

Five women from different corners of the globe came together in Costa Rica's tropical rainforest in 2017. Each was willing to brave long nights, constant rainfall, mudslides, isolation, extreme humidity, and nightmarish insects to volunteer for six months in a novel research and conservation program—determining how to rehabilitate wild sloths and return them to their native tropical forests. Of course, access to wild-grown mangos and starfruit, nearby beaches lined with coconut trees, daily sightings of amazing animals—like toucans, scarlet macaws, and white-faced capuchin monkeys—and developing friendships with a sisterhood of female biologists made the challenges easier to bear.

Our mission was to better understand and conserve two sloth species—the Hoffman's two-toed and the brown-throated three-toed. The brown-throated sloth, a smaller species with a distinctive eye mask, is diurnal (active during daylight) and feeds on leaves in the forest canopy, whereas the much bigger Hoffman's sloth, with a "piggy" nose and longer, lighter fur, is nocturnal. The biggest threats to sloths in Central America are their removal from the wild for the illegal pet trade and the destruction of their natural homes by deforestation.

Displaced sloths can end up isolated in urban areas, unable to return to the unfragmented forest areas on which they rely for their survival. Forest canopy gaps larger than a sloth's arm span prevent the animals from moving through the branches, requiring them to descend to the ground, where they face a variety of often fatal threats. Sloths also sometimes encounter uninsulated power lines, receiving electric shocks.

Our team rescued sloths from these precarious situations and delivered them to veterinary clinics, where they were nursed back to health. We then fitted the sloths with radio collars and gradually reintroduced them into forested areas where they assimilated back into the wild. Once the sloths were released, we tracked their movements through the forest canopy, collecting detailed information about each individual's activities to learn how the reintroduced animals behaved in the wild. We also compared their behavior to that of sloths that had always lived in the wild to determine whether the released sloths were thriving in their natural habitat.

The sloth monitoring team included eight or so researchers housed either at our base camp on the edge of the forest or in a basic village apartment in the small coastal tourist town of Manuel Antonio, about a thirty-minute walk from camp. Team members arrived throughout the year for a six-month stint, and luck of the draw determined where each of us stayed. Each living situation had its challenges. Base camp consisted of four conjoined tents and a kitchen, all located on an elevated platform six feet above the forest floor, and a nearby outdoor bathroom. Sheltered only by a tin roof and canvas walls, the camp was open to the elements and was often visited by forest creatures. The village apartment consisted of a series of concrete-walled, garden-level rooms that flooded during heavy rains. Due to the high humidity, everything (think walls, furniture, bedding) was covered with mold, and friendly stray animals often visited during the day.

Most of our work took place at night, when the Hoffman's two-toed sloths—the most common sloths we released—were active. We each radio-tracked and monitored the one sloth assigned to us each night, taking detailed notes on its location and behavior. We also monitored still-captive sloths to determine whether they were ready to return to the forest. Before their release, these sloths acclimated to their new home by spending time in a large prerelease pen located in the forest and surrounded by large trees.

Our sloth-stalking efforts and rainforest life were marked by hilarity, chaos, and unforgettable moments. And through our shared experiences, we learned valuable lessons about ourselves and the tropical forest—and became lifelong friends.

Coming from far-flung locations and having very different backgrounds, we arrived weeks apart for our respective start dates in verdant Costa Rica after varied and challenging journeys.

Holly had flown for more than thirty-two hours from the small city of Perth in Western Australia. She'd previously lived and worked in a Mediterranean climate amid muted silvery and dark-green woodlands filled with gum trees, spiky dry vegetation, and wild marsupials. Having studied honey possums—tiny, thumb-sized marsupials with a long prehensile tail and a nectar diet—in the Australian bush, where everything seemed designed to kill you, she assumed that working with sloths—which were bigger, slower, and easier to spot—would be a breeze.

Masha hailed from Washington, DC, and was excited to have escaped the traditional nine-to-five job she had held for six months. She dreamed of being alone in the tropical rainforest with only sloths for companions. She, too, had had a trying journey, almost crashing in a small airplane en route and arriving at her new home only thanks to the kindness of strangers and her first hitchhiking experience.

Bruna came from São Paulo, Brazil. After two years spent studying large mammals in an agricultural landscape, she looked forward to working in a different country and ecosystem. What she hadn't bargained for was the heat—and the humidity—and the rain, the endless rain. She'd somehow missed that she might be taking freezing cold showers and living out of a tent—not exactly the place to avail herself of the nail polish and dresses she'd brought for special occasions.

Taylor left her home in Wisconsin in the US eager to explore a new ecosystem and career path after working in a soil and water chemistry lab. Living and working in the wild was going to be way outside her comfort zone, but she was ready to embrace a challenge.

Kate arrived in Costa Rica after a circuitous journey that involved studying bats for eight months in Malawi, East Africa. She'd long been known as a bat girl—or chiropterologist—but had always had a soft spot for arboreal mammals. She'd received her master's degree and contemplated accepting paid work "like a sensible person," but somehow found herself picking up her backpack again (which she admits she had never unpacked).

Despite arriving tired and hot, all five of us marveled at Costa Rica's green, forested beauty. The sloths we came to study were cute and funny, but monitoring these enchanting, slow-moving creatures for nine hours every night was no easy feat. The work required patience and a good eye. We discovered that entertainment helped us through the long, and sometimes tedious, nights: a good book, podcasts or music, and most of all, good friends to share it all with (especially during thunderstorms). We also learned to expect the unexpected, encountered a host of not-exactly-friendly creatures, and devised ways to celebrate both our friendships and our work.

Each volunteer monitored sloths six days—or rather nights—a week, following them in the rain that usually fell during a large portion of our shift. Aside from trying to stay awake and avoid losing our boots to the sticky mud that coated the forest floor, we faced a daunting parade of organisms as the tropical rainforest came awake at night. During our 9 p.m. to 6 a.m. shifts, each hour seemed to be dominated by a new set of creatures.

We usually encountered biting fire ants as we began our night's work. In the next hour, gnats swarmed our headlamps, which made looking straight up to find sloths in the canopy particularly challenging since the tiny insects inevitably flew right into our eyes. Around 11 p.m., we faced rainforest inhabitants that seemed intent on crawling on or biting us: spiders, snakes, and scorpions (oh my!). Wearing headlamps also made us popular targets for the massive flying grasshoppers that emerged between midnight and 1 a.m. They flew directly at our faces for at least an hour while we winced and recoiled from each hit.

Despite being nocturnal, two-toed sloths sleep on and off for fifteen to twenty hours during the course of a day and night. They typically napped between 1:30 and 2:30 a.m., and though we struggled to stay awake during this hardest part of the night shift, we had to stay on our guard since ocelots sometimes came out to play. We usually got a second wind around 2:30

a.m., and this was a good time, if we were lucky, to spot a tamandua—a small anteater with a long snout and a prehensile tail. Around 3 a.m., bats became particularly active, swooping in to eat the many flying insects that were attracted to our headlamps and getting a little too close for comfort at times.

Dawn unleashed the spawn of Satan, which we knew only as the brown bugs of death. These cicadas relentlessly swarmed our faces for the last hour before sunrise. By 4 a.m., the nocturnal animals had started to disappear, and the howler monkeys greeted the day with a chorus of astonishing whooping calls that could be heard more than three miles away. And finally, by 5 a.m. the capuchin monkeys were swinging about, birds were singing, agoutis were foraging, and the diurnal brown-throated three-toed sloths were awakening.

Because we went where our assigned sloths went, our paths crossed only if those of our sloths did. So for most of the night, we were alone with our thoughts . . . and whatever else was out there. We were armed with mace, and if we were worried about our safety, we could also retrieve one of the Super Soaker water guns stored in the release pen. Whenever we heard a loud crashing sound nearby, our instincts told us to head for safety—or at least to grab a Super Soaker. But we soon learned that going to investigate the source of the noise might be a better alternative.

One night, Holly was following a newly released sloth and struggling to keep track of it as it traversed the high canopy. Suddenly, she heard a loud crash. She ignored the sound for a moment while continuing to record data, but then curiosity got the better of her. Heading toward the source of the noise, she soon found a juvenile Hoffman's two-toed sloth that had fallen out of a tree and landed on the forest floor. Holly quickly radioed one of her colleagues who was working nearby, then rushed to retrieve the baby sloth and put him in a carry case so they could take him to base camp for a checkup. A close examination revealed that after being weaned too early, the youngster appeared to have a viral infection that had likely weakened him and caused his fall. The sloth was taken to the rehabilitation center, where the staff could help with his recovery. A few months later, we happily released him to the area where Holly had found him.

The young sloth was not the only creature that fell from the sky on Holly's watch. One night while she was watching a sloth sleep and recording notes, something suddenly dropped to the ground at her feet. Wide-eyed, she realized it was a little tent-making bat, which immediately scrambled around searching for some low-lying vegetation from which it could hang upside down. Tent-making bats create roosting spots for themselves by biting leaves and shaping them into tiny shelters. Sometimes entire colonies huddle together under these makeshift "tents."

After watching it for a while, Holly saw that it was struggling, and radioed some biologists back at base camp for help. When her colleagues arrived, she helped pick up the bat, put it safely into a dark box, then transport it back to camp. Since Holly had been immunized against rabies, she was able to hold the bat with thick gloves and feed it fruit and sugar water while the group waited for a veterinarian to examine it. The little bat, which Holly and her teammates named Dragon, was diagnosed with a broken wing and taken to a rehabilitation facility for surgery. And so Holly learned that loud crashes and objects that fell from the sky at night were not always scary and were sometimes even adorable.

But not all of our nocturnal encounters were so innocuous. Within Masha's first week of work, several of us spotted an ocelot, a thirty-pound tropical rainforest cat that can prey on sloths. One night while Masha and Bruna were with another researcher, they came face-to-face with an ocelot standing next to the trail right where they had tracked one of the radio-collared sloths. The spotted cat stared at them as though they might be its next meal. Ocelots are territorial, and this one seemed to be defending its hunting area. Masha and Bruna quickly looked to their more experienced colleague, a six-foot-five man, for direction. He immediately began clapping and raucously yelling. Masha and Bruna followed his lead, but the ocelot remained in the undergrowth, watching. It wasn't until all three of them blasted music from their phones that the ocelot vanished into the darkness.

After that night, we saw the ocelot periodically. And though we armed ourselves with Super Soakers filled with a water-vinegar mixture to help deter any wild cats that came too close, we couldn't help but feel we were being stalked in the darkness each night.

Other beasts also lurked nearby as we pursued our sloths in the dark forest. One night, while Masha was watching Kermie—the first sloth to be rehabilitated and released to the wild—feeding, she heard rustling leaves on the ground near her feet. Looking down, she saw a monstrous tarantula. Riveted, she shined her headlamp on the intimidating arachnid. There was no way she was going to let it out of her sight, in case it tried to crawl on her while she kept watch over Kermie. Slowly, it moved toward her. Then suddenly, she felt a slight breeze waft across her face as a bat swooped in, grabbed the spider, and flew away. Masha sagged in relief and shock. You never knew what you might witness in the jungle. Or what creatures might lie in wait for you.

One early morning, around 2 a.m., while Holly was watching a sloth sleep, she placed her rain poncho on the ground so she could sit and rest her legs for a while. Sitting on the forest floor was always a gamble; you never knew which creepy-crawly might join you. As she recorded notes, she felt something inside her sleeve brush against her wrist. Used to chasing away mosquitoes and other pesky insects that hovered around her headlamp, she absently swatted at the spot. A sharp, stinging pain followed, and she looked down to find a scorpion the size of her hand sitting in her sleeve.

Freaking out, she shook her arm and then realized she couldn't see where it had landed—until she spotted it on her leg. Jumping to her feet, she swept it away, making sure the scorpion landed on the ground this time. Having no idea how she would react to the sting, Holly became concerned when her lips began to tingle. Soon the tingling spread until she had a strong pins-and-needles sensation all over her body, so she hiked back up the mountain and all the way to her village apartment, where nurse Masha offered medicine and emotional support.

Just a few days later, Kate found herself in a similar situation. While searching for a sloth after a rainstorm, she grabbed onto a small tree trunk to avoid slipping on the mud, and felt a sharp sting on her hand. Seconds later, a tiny scorpion crawled around the tree to meet her. After about an hour, she felt tingly all over, just as Holly had. Each had had a painful reminder: When you're working in the tropics, check before putting your hands anywhere!

Fortunately, not all of the crew's tropical encounters were so uncomfortable. Another night, Masha was observing Kermie when she noticed a frog sitting on a wooden post nearby. The rain was falling steadily, and she only noticed the little amphibian when she heard a *ribbit*. The frog seemed equally oblivious of Masha until she shined her headlamp on it. Turning toward her, the frog tensed, then lunged straight for her face—and seemed to disappear.

Jumping in surprise, Masha ran over to Taylor, who was monitoring another sloth nearby, and asked, "Can you check my poncho for a frog? I didn't see where it landed after it jumped toward my face." With a laugh, Taylor pointed to the frog that was sitting on Masha's shoulder. Carefully, Taylor helped the frog to a nearby tree, where it resumed its nightly activities.

Unexpected encounters—and not always friendly ones—also occurred during the day. One morning, while Masha was tracking a wild three-toed sloth, a monkey *threw a mango at her head*! (Yes, it's okay to laugh.) Luckily, the mango missed her face by a few inches, but the white-faced capuchin's toothy grin gave the impression that the unrepentant monkey was giggling.

Not long after, Masha caught the same capuchin rummaging through her bag. Before she could react, it was running off with her iPhone. It stopped to stare at her and she stared back, contemplating how to retrieve her phone. Spotting a mango a few feet away, Masha slowly reached out, grabbed it, and rolled it over to the monkey in an attempt at peaceful diplomatic negotiations. The capuchin considered the offer for a moment, then finding the bargain irresistible, dropped the phone, grabbed the mango, and ran away.

Despite their mischievousness, the monkeys were a treat for all of us. It was exciting to see red-backed squirrel monkeys—Costa Rica's most endangered primate—dance about the canopy, their calls sounding like chirping birdsong. The white-faced capuchins were also charming, particularly when the little babies clung to their mothers' backs as the monkeys traveled through the canopy. But sometimes our naivete about the primates got us into trouble.

One early morning just after sunrise, Holly and Taylor had finished their shift and were returning their tracking equipment to base camp when a group of capuchins passed by. Some of the monkeys were at face height in the trees. The two women watched the primates excitedly, pointing out one of them, the patriarch of the group, that looked like an old man. Grinning at the surrounding monkeys, they were unaware that their own bared-teeth expressions posed an inadvertent challenge to the old man—and that he accepted their challenge.

Leaving base camp, Holly and Taylor walked up the road toward their apartment. But hearing a noise behind them, they glanced back to see that the old man and a horde of his friends had left the trees and were now pursuing them. Delirious after working all night, the women burst out laughing at the hilarious scene. Although the capuchins were gaining on them, they somehow managed to outrun the monkeys, which finally gave up their pursuit. After that, we always remembered monkey etiquette and refrained from ever grinning at monkeys again. And from then on, we averted our gazes whenever we saw the old man—even when the impish monkeys convened on the tin roof of the base camp kitchen, keeping Bruna awake with their riotous play.

At times, the elements could be as trying as the fauna in Costa Rica. The ground-level apartment that housed Taylor, Masha, and Holly was without a front door for an extended period and did little to keep moisture at bay. They had a shower they couldn't use because of the constant presence of worms. On most days, a dehumidifier worked well enough to minimize the moisture, and they could wipe down surfaces with vinegar to kill the mold for a few days, but the trio was utterly foiled by heavy rainfall. One night, they returned home at 3 a.m., eager to rip off their muddy boots, change their sweaty clothes, and collapse into bed. Instead, they were met with ankle-deep water in the apartment. For a couple of hours they took turns using their only two towels to sop up the water, wringing them out into a bucket. They even tried to use condiment bottles the way you would use a turkey baster to draw up water and squirt it into the bucket, laughing at their ridiculous situation while working diligently as the rain kept falling outside.

Torrential rain posed innumerable difficulties for our work as well. Following and observing our sloths was challenging since we had to navigate around obstacles on the ground while our subjects roamed freely in the canopy, but at least they were slow—as their name suggests. When it rains, though, sloths suddenly shift gears and can actually disappear quite quickly in what we dubbed the Sloth Rain Dance. This unexpected behavior was especially problematic when the rain turned into a storm, and we saw some awesome storms. Torrential rains that quickly turned forest paths into cascading waterfalls, deafening thunder claps, and lightning that seemed to strike the ground—scary stuff. Nonetheless, we all tried our best to stick to our posts, squinting up into the canopy as rain fell into our eyes, and trying and failing to relocate our sloths as they swiftly vanished.

But even the most determined field biologist has her limits. To better cope with our challenging conditions, we decided to create a makeshift shelter using a tarp and sticks so that when the rain fell too heavily for us to do our work, those who were close by could retreat to the shelter to wait out the storm. Sometimes, torrential rains could last for hours—and that's what precipitated the first of our many tropical rainforest dance parties. Kate, Masha, and Bruna met under the tarp (thank you, walkie-talkies!) and played music to counteract the shivers. Soon they were dancing and singing at the top of their lungs, like no one was watching, all dressed up in their Sunday best—boots and raingear—with headlamps used as disco lighting. The pounding rain drowned out their voices but couldn't stop their uproarious laughter.

It was often the little things that got us through our long night shifts: listening to a podcast, anticipating a snack awaiting us in our tent or apartment, and the prospect of a shower, clean clothes, or sleep. But sometimes the end of our shifts promised something more exciting. One night, Kate and Taylor decided to celebrate the end of their graveyard shift with an early morning swim in the Pacific Ocean. The off-limits private beach they planned to visit was picturesque, often vacant early in the morning, and less than a five-minute walk from our field sites. They were nearing the end of their volunteer stints, and their deference to the private beach restrictions had eroded after listening to the waves call to them night after night. Eager to embrace the temptation at last, they wore swimsuits under their field clothes and readied themselves for a final, blissful, celebratory swim.

Their anticipation built all night, brightening even the usually dark period when the sloths were sleeping. At long last, their shift ended. Kate and Taylor stepped into the tranquil ocean, then floated under dawn's pastel skies, witnessed only by a dispersing crowd of hermit crabs. They went to bed after the sun rose but slept more soundly in their stifling quarters than they had since they'd arrived.

By the time our work commitments ended, we five had become fierce and fearless friends. Each of us left Costa Rica with notes from the others reflecting on the time we had spent together. Reading our friends' words on our way back to civilization allowed us to reminisce about the adventures we'd shared, reflect on the depth of the bonds we'd formed, and take pride in what each of us had accomplished.

Despite the challenging working and living conditions we endured, we all look back nostalgically on what we experienced together: the shared secrets, the singing, the laughs, the captivating wildlife, the long nights in the rain-soaked forest, and the bright days by golden beaches. Above all, the support and confidence we gave each other was like getting the "right shoes," empowering us to handle future work and chase our next adventures. Together, we'd weathered the jungle—where only the strongest survive.

Amberley Bradley

Holly Bradley is a conservation biologist in Perth, Western Australia, working toward the conservation and protection of several of Australia's unique and wacky threatened fauna, including the spiny-tailed skink and the ghost bat. She has a doctorate in threatened species ecology and conservation management, and spends her time surrounded by the forests and expansive beaches of the West Aussie southwest or flying to the red dirt and sweltering heat of the Pilbara.

Scott Taylor

Masha Edmondson is a biologist with a passion for wildlife who bridges the gap between research and policy to find sustainable solutions to environmental challenges. She grew up in Washington, DC, earned a master's in coastal environmental management from Duke University, and has conducted research in a range of locations from Greenland's glaciers to Costa Rica's tropical forests. She treasures her fearless field biologist friends—and is inspired by the fierce women paving the way in STEM fields.

Tiago V. Fernandes

Bruna Campos is a Brazilian biologist with a master's in ecology and conservation. She works remotely as an environmental consultant while living in Europe, blending her career with a love for new landscapes. After volunteering in Costa Rica, she really misses the smells, sounds, and beauty of that country, and most of all being close to, and sharing adventures with, the amazing women she worked with there.

Courtesy of Taylor McDowell

Taylor McDowell grew up in Hartland, Wisconsin, where she spent most of her time playing with her pets or exploring the outdoors with her sister, early experiences that fostered a love of animals and the outdoors. She worked briefly in wildlife conservation but ultimately earned her PhD in geography from the University of Wisconsin Madison, where she studied how soils and landscapes change over time. Today, she continues to research both natural and human-modified soils and landscapes and loves observing wildlife both on hikes and in the neighborhood.

Courtesy of Kate Derrick

Kate Derrick is originally from the United Kingdom but has lived in France since she was twelve. After receiving her master's in wildlife management, she worked with bats in France, Africa, and Dominica, and with sloths in Costa Rica. She lives in Corsica and works as a researcher with the Corsica Bat Group. An avid music fan, she loves going to festivals and metal concerts and sings in a band. She also plays roller derby and loves to draw.

The Good Life

Maddie Halloran

I stood on the edge of a big, smooth rock, staring across a long, deep pool of the clearest, greenest water I had ever seen. The Smith River is the longest wild and scenic river in California, and swimming in it has been one of the great privileges of my life. But whenever I readied myself for any river dive, I always had to take a slow, focused breath before pulling on my wet suit hood. Something about having my ears encased in neoprene while out of the water always made me feel claustrophobic—and extremely vulnerable.

As I took my centering breath, one of my fellow dive-team members turned back to me before he jumped into the water and asked, "Hey Maddie, what is a good life?"

"What?" I said, startled.

"I'm just wondering. What do you think makes a good life? I'm about to turn thirty, and I need some answers, so I've just started asking people this question."

Perplexed, but mildly flattered that anyone would think I knew anything about living a good life, particularly as a twenty-six-year-old graduate student, I shook my head.

"Give me a minute to think about it. I'll tell you at the end of the dive." And with that, I pulled on my hood and jumped in.

The Smith River is so far north in California that its northern fork actually starts in Oregon. Every summer for the last thirty years, a group of fish enthusiasts has gathered to camp for a weekend and conduct a snorkel survey of several key species—Chinook salmon, summer steelhead, and cutthroat trout—that call this watershed home. This type of long-running monitoring helps to assess the effectiveness of restoration and protection efforts for at-risk fish species. A mix of hardy locals and biologists from all over the state volunteer to spend a Friday afternoon getting trained to identify and count the fish underwater. Then on Saturday we split up, and each team is assigned a several-mile-long stretch of the river, where we snorkel downstream and count the fish we see.

I'd originally learned to conduct snorkel surveys for coho salmon two years earlier on the creeks surrounding Point Reyes National Seashore, just north of San Francisco. These watersheds are incredibly unique and beautiful, but in the summer, when our field crew began to snorkel, the streams were in their most shrunken state. Many of the pools we surveyed for juvenile salmon were not even knee-deep, and most of the fish we saw were the length of my index finger.

I didn't understand why people got so excited about doing snorkel surveys, because mostly what I saw was shallow brown muck swirling in front of my face as I army-crawled my way upstream. This was nothing like the tropical snorkeling I had done on trips to Hawaii or French Polynesia and was something I would only ever do if someone were paying me.

Luckily, my coworker that first summer convinced me that we should volunteer for the dive weekend on the Smith River, a seven-hour drive north but a world away from the shallow Point Reyes streams. The Smith River watershed differs drastically from many of the state's more populated areas farther south. Enormous redwoods tower over winding single-lane roads, and even on sunny summer days the water can be brutally cold. Depending on the stretch we're assigned to survey, we could be out snorkeling for up to five hours, so we wear neoprene gloves and hoods along with full-body wet suits that provide both warmth and buoyancy.

We work a reach from upstream to downstream, so although it's called a dive, it's really a series of long, slow drifts through deep pools, broken up by body surfing over fast, shallow riffles. Every hundred yards or so, each team of snorkelers gathers and gives their group leader a count of how many fish they saw on that stretch, and in this manner, everyone works their way down their section of river.

The first time I snorkeled in the Smith, I slowly maneuvered my body into the water and rode a small riffle into our first survey pool. Instantly, the bottom dropped away. The pool was at least thirty feet deep, the water clear enough that I could see straight to the gravel bottom where a dozen cutthroat trout, each easily as long as my forearm, moved in sync. This was completely different from looking for fish in Point Reyes streams, but it wasn't anything like the tropical snorkeling I had done either—it was somehow even more magical.

Conducting my graduate research on coho salmon had given me a new appreciation for the seasonal rhythms that unfold in West Coast rivers. All five species of Pacific salmon are anadromous; they hatch in freshwater streams and remain there for varying amounts of time before migrating to the ocean, where they grow to maturity. The adults return to their natal stream to spawn and then die. Because it's difficult and expensive to track fish in the ocean, salmon monitoring projects usually take place in the rivers and streams where the adults spawn and the juvenile fish grow after hatching. In the summer, those of us who study salmon snorkel sections of rivers and streams to count the fish that hatched during the previous winter. In the fall, we hike up streams looking for spawning adults and their redds—the gravel nests that salmon create to hold their eggs. In the spring, we install smolt traps to catch the hand-length yearling fish as they migrate to the ocean. Through this seasonal work, I moved through the calendar year in sync with the salmon life cycle, aware of patterns, rhythms, and seasons I otherwise might not have noticed.

In between dodging boulders and identifying the quicksilver fish during my second summer of diving the Smith, I thought about my team member's question: What *is* a good life? I wasn't sure I knew, but I was grateful to be asked anything difficult because it distracted me from the stress of leading a dive for the first time. I wasn't sure why I was so nervous. I was with two of my good friends, and we were all experienced snorkelers. I later realized that it was precisely *because* I was in charge of my friends' well-being that I was so anxious. I not only wanted to make sure that we all had an amazing day, but I also needed to ensure that everyone successfully navigated the river's hazards and got back to camp safely.

I have experienced many flavors of anxiety during fieldwork. When I describe my previous jobs—measuring plants in beautiful national parks, studying sediments on the Swedish tundra, counting fish in incredible rivers—to friends and family, fieldwork sounds fun and fascinating and sometimes almost glamorous. But it can also be dangerous and wild. I have fallen down scree slopes, flooded my waders in rushing waters, stared down a mountain lion from forty yards away, and nearly fainted from heat exhaustion.

Working far from home in remote areas often adds to my unease, and I frequently feel scared before, during, and even after time in the field. Sometimes just the anticipation of what I might encounter on any given day is enough to put me on edge. That fear and anxiety has caused me to cry in frustration while camping in absurdly beautiful places and shout at the plants I was supposed to be counting.

That day on the Smith River, though, the dive went perfectly. Once we were all safely back at the campsite with our first round of beers open and a potluck dinner imminent, I basked in a feeling of deep satisfaction. What a wildly incredible way to spend my time. As I thought about my day—the emerald water, the fascinating fish, the camaraderie of doing what I loved with friends—it occurred to me that this whole weekend was the good life. Maybe I'd been living the good life all along. I'd just never stopped to think about it—no one had ever asked me before.

That summer, for the first time in years, I felt nearly settled. I had a group of friends I loved, and I was in constant awe of wild salmon and of how much I still enjoyed learning about their secret lives, day after day. It had been an immensely frustrating and difficult journey to reach this place of satisfaction. I had spent years watching my friends confidently choose careers, jobs, and places to live; sometimes it felt like everyone had a plan except for me. But this incredible fisheries work—and all the great biologists I'd come to know while doing it—was worth the weird twists and turns my life had taken, and the many years of uncertainty during which I'd struggled to decide if I wanted to be a scientist at all, let alone what kind.

Looking back, I realized that if I hadn't persisted through the mistakes and missteps I'd made along the way, I might not have ended up counting these impressive fish on the Smith River in the sunshine of a perfect July day. It made sense that I needed to work for a consulting firm to decide that I could never be an engineer. I had to spend a field season in the desert to realize that what I love most is being cold and wet.

Later that evening, while sitting around the campfire searching the sky for shooting stars, I shared my answer with my friend. For me, the good life is simply the satisfaction that comes from being in a beautiful place, doing meaningful work with great people. After leading the day's successful dive, I realized, too, that I am always most satisfied when I can make sure that the people around me are safe and happy.

Being a human is hard. Many of us are plagued by self-doubt and indecision, consumed by our daily struggles and insecurities. Too often, we agonize over what we think we should be doing or achieving. But that day on the Smith, I realized that it's all the good life. *All of it.* It's important sometimes to pause and appreciate that. And when you look back on the journey you've taken, who knows? It may seem like you had a plan all along.

Marg Halloran

Maddie Halloran was born and raised on Dena'ina lands in Anchorage, Alaska, and has been lucky enough to travel, work, and play in special places all over her home state. After spending summers in high school and college working in tourism, she earned a bachelor of arts in environmental studies from Carleton College. Since then, she has worked in various environmental science and education positions in Alaska, Maine, Minnesota, Arizona, and California.

Maddie earned a master's in fisheries biology at Humboldt State University, where her research focused on local population dynamics of coho salmon. After spending years learning about the difficult road to salmon restoration in places harmed by decades of environmental destruction, she moved back to Anchorage to work for a nonprofit organization that seeks to protect public lands and waters in Alaska. Her hobbies include cross-country skiing, ultimate Frisbee, knitting, and taking long walks with her dog.

How to Grow Wild Friendships

Kira A. Cassidy

Fieldwork provides the perfect environment for growing close friendships. Before I started working as a field biologist, I sometimes found it difficult to create deep connections with others. While I enjoyed meeting and interacting with people, getting to know someone initially was often nerve-racking. Parties and large group gatherings could be fun and I usually left them smiling, but I sometimes felt drained after struggling to make conversation or find things in common with people I didn't know very well.

On the other hand, there are a handful of people who I felt like I'd known for a long time from the first time we met. We somehow skipped through the disjointed small talk that's typical between strangers and launched right into a friendship that was easy and comfortable. We soon had a bond with depth, one that was an unusual combination of calming and energizing. We talked about important topics and seemed to understand each other intuitively. These rare connections were almost always rooted in wildlife fieldwork.

Wildlife work usually involves collecting scientific data and being outside—two of my favorite things. When working in the field, you are often paired with someone or assigned to a small group to achieve a common purpose. You have to work together to solve problems, and you share a wide range of experiences. Together you navigate flat tires in the work truck, endure thunderstorms and mosquitoes, brave river crossings and obstinate bison blocking the trail, and stay up late to fill out data forms and label samples. You relish time outdoors and delight in endless unexpected discoveries. You might hike to a place where a secretive mountain lion cached its mule deer kill weeks ago. Or watch ground squirrels calling in alarm as a badger hunts nearby, or witness a cow moose giving birth. You spend hours together each day and quickly get to know each other's life stories. If you're lucky, some of your teammates become your lifelong friends.

I met Erin during my third field assignment, my first in Yellowstone National Park. We've now known each other and worked together for more than eighteen years. We initially spent thirty days tracking the Agate Creek wolf pack. We hiked or snowshoed to their elk kills, where we collected samples—a tooth for age estimation, bone marrow to assess fat reserves—and recorded data that would provide insights into the wolves' lives. When the pack was out of sight, we chatted and exchanged books to pass the time. Since then, we've searched for wolf dens (sometimes taking a detour to visit a raspberry patch), necropsied animals to determine why they died, and pored over countless data forms.

Each year, Erin and I don flight suits and helmets and climb into a helicopter to capture and radio-collar wolves in Yellowstone. We move efficiently, knowing instinctively what the other will do, even during the high-stress work of animal capture. We take turns holding a wolf's front leg while the other person carefully draws blood samples.

On one scary occasion, when a wolf's internal temperature swung from too high to too low after the animal was immobilized with a tranquilizer dart, Erin was the one person on the planet I wanted—and needed—to have there with me. The wolf vomited a few times and then stopped breathing. In a split second, Erin and I launched into emergency mode. I cleared the wolf's airway and positioned her to give the Heimlich maneuver and possibly CPR. Erin pinched the wolf's face, nose, and ribs, the equivalent of a sternal rub on a human—a pain test EMTs use to assess brain function. After the wolf started breathing again, we stayed with her for several hours to make sure she was okay. Later, with dried wolf saliva and vomit covering our gloves and sleeves, Erin and I hugged and thanked each other, grateful for one another's steadying presence.

Outside of work, we have set up craft booths at the local fair, gathered to watch every episode of *Game of Thrones*, and celebrated every Thanksgiving together. I even held her son when he was only twenty-four hours old.

My friend Lisa is like a big sister to me. We met in Yellowstone, where Lisa and her husband anchored a badass crew of backcountry skiers who collected data from wolf-killed carcasses for thirty days straight each winter. For fifteen years, Lisa and I occasionally worked on the same crew, and I loved chatting as we hiked, or glancing up from my spotting scope to see her expression mirroring my own—usually a shocked smile upon witnessing some remarkable wolf behavior.

I'll always cherish the time we stood in the dark outside our bunkhouse, frozen in wonder, listening to the Blacktail pack howl, pant, and run through the grass just out of our view in the dim moonlight. Or when we watched in delight as a weasel curiously approached us, a bright-white little creature hopping around Douglas-fir roots. Or the time we skied to a backcountry cabin, stopping along the way to collect moose pellets and urine samples so we could identify individual animals and get more accurate counts of the park's moose population. We laughed until our sides hurt as we shared stories, and once we were tucked into our cabin for the evening, we competed fiercely at board games. We necropsied dozens of animals together—assignments that could be emotionally taxing, especially when it was a wolf we'd watched for many years.

I was always glad for Lisa's companionship on the fun days, and even more grateful on the tough days. We have always supported each other—whether navigating through a tricky stretch of fallen trees or struggling with tough life decisions. Even though we live more than two thousand miles apart for most of the year, I always look forward to getting one of Lisa's big hugs once we are together again.

Cheyenne and I were dropped off in a remote part of Ellesmere Island in far northern Nunavut, Canada, to observe a pack of arctic wolves and assess whether they would tolerate a documentary crew filming them. We had worked together on several occasions in the US and knew each other well, but our shared experience just a few hundred miles from the North Pole provided me with one of my favorite field memories.

Around midnight, in the twenty-four-hour polar sunlight, two of the younger wolves started to approach our camp but then turned back toward their den about a mile away. We didn't want to frighten them but needed to see if they would accept the close presence of filmmakers. Rather than risk intimidating them by approaching, Cheyenne and I started playing tag with each other, hoping to arouse the wolves' natural curiosity. When I chanced a quick glance at them out of the corner of my eye, I saw the wolves watching us—ears perked, heads tilting inquisitively. We'd piqued their interest, and they cautiously moved toward us again. When we stopped playing and breathlessly looked at them to gauge their reactions, they turned away. We resumed playing until eventually our curious canine neighbors came within about a hundred feet.

No longer seeming confused by our actions, they now appeared to want to join in the game. The young wolves started playing and wrestling with each other nearby, pausing to glance over at *us* occasionally, almost as if checking to see if we were watching them play, too. Cheyenne and I still reminisce about that day more than a decade ago, relishing that unforgettable sunlit night when we played tag with wolves—they in their game and we in ours, side by side.

These women and other field partners who have become my close friends have shared all kinds of unique experiences with me: ones that have been happy, funny, chaotic, stressful, heartbreaking. Together we've watched sunrises, identified wildflowers, found perfect places to camp, cooked delicious backcountry meals, and listened to owls and wolves calling at night. We've pondered the complex lives of the wild creatures we've studied, and reveled in the wild places where we've worked—areas rarely seen by other people. We've endured pounding hail, sweltering heat, frightening storms, heavy backpacks, and broken equipment. We've processed the bodies—and emotionally processed the deaths—of favorite longtime study animals. Together.

After a fieldwork assignment is over, our individual lives continue to unfold, and in some cases, we may not see each other again for weeks, months, even years. Yet when we reunite, we easily find our rhythm. We fondly remember our first meeting and the many special moments we've experienced together since then. Our lasting bonds are nourished by shared adventures. As I and so many other field biologists have been lucky enough to discover, nature is the perfect substrate for growing wild friendships.

Acknowledgments

We thank every biologist, friend, and family member with whom we've discussed this project for the universally enthusiastic and encouraging responses. You're the best! This group includes Norah Bruce, who, through her youthful exuberance for all things nature and her eagerness to explore any and all streams, was Susie's initial inspiration for this anthology.

Several people read one or more story drafts and offered feedback on them, and for that we thank Sophie Quinn, Norah Bruce, Elaine Gelbard, and Joanne Wilkinson. Mark Wilkinson provided expert technical review for one story. Kate Lechler provided early edits of "Stalker." Donna Levine replied to our grammar questions and, unbeknownst to her, was the final arbiter of our occasional comma impasses. Meridith Wulff was always generous with her librarian smarts. Sophie Q and Norah B served as our Gen Z cultural consultants when we needed to understand what "kids these days" know (for example, *Do you know who Yosemite Sam is?*). Matthew Derr and Karin Snelson read and gave helpful feedback on some of our submission materials for potential publishers.

We are profoundly grateful to and impressed by the many women biologists who submitted stories to this anthology project. It takes bravery and significant effort to step out of scientist boots and into the shoes of a narrative nonfiction writer. Your vulnerability in sharing your stories was admirable. *All* of your stories are wonderful and important, and we dearly wish we could have included every one!

Special thanks to our contributing authors. You displayed incredible forbearance and dedication in working with us through multiple rounds of intensive editing and story-crafting. We hope you are as excited about your final stories as we are. And we hope to meet each of you in person one day!

We thank the team at Mountaineers Books for bringing this anthology to fruition. Kate Rogers and Emily White saw the value in the project from the outset. Others who were instrumental include project manager Laura Shauger, designer Jen Grable, editorial assistant Natalie Mannino, proofer Callie Stoker-Graham, and publicist Kate Jay. Lorraine Anderson is a phenomenal editor who added the final polish to many of the stories with a light but insightful touch.

I (Susie) thank my family for their patience and support in many ways, especially Charles for keeping me fed, Casey for being a sounding board, and Sophie Q for *always* offering encouragement and believing in her mom. Thanks to Eleanor and John Green for graciously hosting me at the Green Farm during intensive book proposal writing and to Susan and Lou Zachos for providing space for my book work just up the road from my regular job.

I am grateful that serendipity led me to team up with Sophie O, who, luckily, turned out to be the perfect coeditor. You taught me so much about crafting stories, being a kind and effusive editor, and never, ever cutting corners. And your patience when the going got tough was the gift that made this book happen. I never could have done it on my own. Plus, I now have a great new friend!

I (Sophie) thank my sisters, Lisa Osborn Koitzsch and Natasha Osborn, for never failing—not once in five years—to ask how my weekly call with Susie went. They have been my superpower through a lifetime of highs and lows, providing unwavering support, steadfast loyalty, and endless laughter.

Reconnecting with Susie was indeed serendipitous. Thank you for asking me to join you in realizing your dream of creating this anthology. With your concise prose, incisive editing, fierce efficiency, and unstinting generosity, you were the ideal coeditor. Together, we became better writers and editors, argued about grammar, struggled to craft perfect sentences, laughed at our foibles, and shared in each other's lives.

Working so closely together, week after week, could have taxed the best of relationships. Instead, it forged two acquaintances into incredible friends. I am so glad to have shared this journey with you!

About the Editors

Bill Dabney Photography

Susan (Susie) B. Adams is an internationally recognized freshwater ecologist and a senior scientist at a federal agency. Her research focuses on crayfish and fish ecology and examines human influences on freshwater ecosystems. Susie has published dozens of scientific papers, presented her research on four continents, and served as president of the International Association of Astacology (study of crayfish). She is committed to helping women find their places in science and the outdoors. Find out what she's up to at adamsecology.net.

Lisa Osborn Koitzsch

Sophie A. H. Osborn is an award-winning writer and wildlife biologist who has worked with birds throughout the Americas. Sophie contributed to reintroduction efforts for several endangered bird species, including as a field manager for the California Condor Recovery Program in Arizona. Her book *Condors in Canyon Country* won the 2007 National Outdoor Book Award for Nature and the Environment. Reviewers for the American Birding Association chose her second book, *Feather Trails: A Journey of Discovery Among Endangered Birds*, as their favorite bird book of 2024. Find out what she's up to at www.wordsforbirds.net.

A portion of the authors' proceeds will be donated to nonprofits that help girls experience science in the outdoors, including the Montana Natural History Center's STEEM Camp for middle school girls and the Great Smoky Mountains Institute's Girls in Science Camp for ages 12 to 15.

ABOUT SKIPSTONE

Skipstone is an imprint of independent, nonprofit publisher Mountaineers Books. It features thematically related titles that promote a deeper connection to our natural world through sustainable practice and backyard activism. Our readers live smart, play well, and typically engage with the community around them. Skipstone titles explore healthy lifestyles and how an outdoor life relates to the well-being of our planet, as well as of our own neighborhoods. Sustainable foods and gardens; healthful living; realistic and doable conservation at home; modern aspirations for community—Skipstone tries to address such topics in ways that emphasize active living, local and grassroots practices, and a small footprint.

Our hope is that Skipstone books will inspire you to effect change without losing your sense of humor, to celebrate the freedom and generosity of a life outdoors, and to move forward with gentle leaps or breathtaking bounds.

All of our publications, as part of our 501(c)(3) nonprofit program, are made possible through the generosity of donors and through sales of 700 titles on outdoor recreation, sustainable lifestyle, and conservation. To donate, purchase books, or learn more, visit us online:

SKIPSTONE

LIVE LIFE

MAKE RIPPLES

www.skipstonebooks.org

www.mountaineersbooks.org